THE ANATOMY and PHYSIOLOGY LEARNING SYSTEM:

Instructor's Manual

Edith J. Applegate, MS

Professor of Science and Mathematics
Kettering College of Medical Arts
Kettering, Ohio

W.B. SAUNDERS COMPANY
A Division of Harcourt Brace & Company
Philadelphia London Toronto Montreal Sydney Tokyo

P9-CRU-905

W.B. SAUNDERS COMPANY
A Division of Harcourt Brace & Company

The Curtis Center
Independence Square West
Philadelphia, Pennsylvania 19106

THE ANATOMY AND PHYSIOLOGY LEARNING SYSTEM:
INSTRUCTOR'S MANUAL

ISBN 0–7216–6636–1

Printed in the United States of America.

Last digit is the print number: 9   8   7   6   5   4   3   2   1

PREFACE

This manual is designed to assist instructors who use *The Anatomy and Physiology Learning System*. It is organized by chapters that correspond to the chapters in the textbook and workbook. Each chapter in the manual has the following components:

- **Key Terms/Concepts**
- **Chapter Objectives**
- **Chapter Outline/Summary**
- **Answers to Review Questions**
- **Answers to Learning Exercises**
- **Answers to Chapter Self-Quiz**
- **Answers to Terminology Exercises**
- **Answers to Fun and Games**
- **Quiz/Test Questions**

The key terms and concepts are defined for quick and easy reference. The chapter objectives are listed in the order in which they are discussed in the textbook. The chapter outline/summary offers a quick review of the important points in the chapter and the summary is referenced to the objectives. The next component, Answers to Review Questions, provides the answers to the review questions at the end of each chapter in the textbook. The next four components provide answers to workbook features. Most of the workbook consists of learning exercises and this manual provides the answers for these. This manual also provides the answers for the chapter self-quiz, terminology exercises, and fun and games that appear with each workbook chapter. Finally, there are 25 quiz/test questions, with answers, for each chapter. Fifteen of the questions are of the type where the student names a structure, feature, or process that is described. Ten of the question are of the true/false type. These are useful for a quick quiz or may be used as part of a larger test.

Although not a part of this manual, there is a computerized test bank available for adopters of *The Anatomy and Physiology Learning System*. The test bank contains 50 multiple-choice questions for each chapter and these are referenced to the objectives for the chapter.

Sometimes I think there are little gremlins that gain pleasure from sneaking errors into answer keys and instructor's manuals. Unfortunately, I am responsible for any errors in this manual. I have no one else to blame. If you find any errors, please let me know so I can correct them for the future.

One of the most rewarding experiences of teaching is working with students on an individual basis or in small groups, helping them understand important concepts. Many of us get so busy that our time for this type of activity is limited. I hope *The Anatomy and Physiology Learning System* in general and this manual in particular will give you some additional time for **real teaching**, the kind that makes a difference in student lives and attitudes. May teaching A&P be an enjoyable and rewarding experience, both for you and your students..

CONTENTS

1 Introduction to Anatomy and Physiology

☞ **Key Terms/Concepts**

Anabolism Phase of metabolism in which energy is used to build complex molecules from smaller ones; synthesis reactions; opposite of catabolism.

Anatomical position Standard reference position for the body; body is erect, facing the observer, upper extremities are at the sides; palms and toes are directed forward.

Anatomy Study of body structure and the relationships of its parts.

Catabolism Phase of metabolism in which complex molecules are broken down into smaller ones and energy is released; opposite of anabolism.

Differentiation Process by which cells become structurally and functionally specialized.

Homeostasis A normal stable condition in which the body's internal environment remains the same; constant internal environment.

Metabolism The total of all biochemical reactions that take place in the body; includes anabolism and catabolism.

Negative feedback A mechanism of response in which a stimulus initiates reactions that reduce the stimulus.

Physiology Study of the functions of living organisms and their parts.

☞ **Chapter Objectives**

Upon completion of this chapter the student should be able to:

1. Define the terms anatomy and physiology and discuss the relationship between the two areas of study.
2. List the six levels of organization within the human body.
3. Name the eleven organ systems of the body and briefly describe the major role of each one.
4. List and define ten life processes in the human body.
5. List five physical environmental factors necessary for survival of the individual.
6. Discuss the concept of homeostasis.
7. Distinguish between negative feedback mechanisms and positive feedback mechanisms.
8. Describe the four criteria that are used to describe the anatomic position.
9. Distinguish between the terms superior and inferior, anterior and posterior, medial and lateral, proximal and distal, superficial and deep, visceral and parietal.
10. Describe sagittal, midsagittal, transverse, and frontal planes.
11. Distinguish between the dorsal body cavity and the ventral body cavity, and list the subdivisions of each one.
12. Locate the nine abdominal regions.
13. Distinguish between the axial and appendicular portions of the body.
14. Use anatomic terms that relate to specific body areas.

☞ **Chapter Outline/Summary**

Anatomy and Physiology (Objective 1)
● Anatomy is the scientific study of structure.
● Physiology is the scientific study of function.
● Anatomy and physiology are interrelated because structure has an effect on function and function influences structure.

Levels of Organization (Objective 2)
● From simplest to most complex, the six levels of organization are chemical, cellular, tissue, organ, body system, and total organism.

Organ Systems (Objective 3)
● The integumentary system consists of the skin with its derivatives; it covers and protects the body.
● The skeletal system includes the bones, cartilages, and ligaments; it forms the framework of the body.
● The muscular system consists of all the muscles in the body; it produces movement and produces heat.
● The nervous system includes the brain, spinal cord, and nerves; it receives/transmits stimuli and coordinates body activities.
● The endocrine system consists of the ductless glands; it regulates metabolic activities.
● The cardiovascular system includes the blood, heart, and blood vessels; it transports substances throughout the body.
● The lymphatic system includes lymph, lymphatic vessels, and lymphoid organs; it is a major defense against disease.

- The digestive system consists of the gastrointestinal tract and accessory organs; it is responsible for the ingestion, digestion, and absorption of food.
- The respiratory system includes the air passageways and lungs; it is responsible for the exchange of gases between the external environment and the blood.
- The urinary system consists of the kidneys, urinary bladder, and ducts; it functions to eliminate metabolic wastes from the body.
- The reproductive system consists of the ovaries and testes with the associated accessory organs; its function is to form new individuals for the continuation of the species.

Life Processes (Objectives 4 and 5)
- The basic characteristics that distinguish living from non-living forms are organization, metabolism, responsiveness, movement, and reproduction.
- Human life has additional characteristics such as growth, development, digestion, respiration, and excretion.
- Physical factors from the environment that are necessary for human life include water, oxygen, nutrients, heat, and pressure.

Homeostasis (Objectives 6 and 7)
- Homeostasis refers to a constant internal environment.
- A lack of homeostasis leads to illness or disease.
- Homeostasis is usually maintained by negative feedback mechanisms, which inhibit changes.
- All organ systems of the body, under direction from the nervous and endocrine systems, work together to maintain homeostasis.
- Positive feedback mechanisms are stimulating and cause a process or change to occur at faster and faster rates.

Anatomical Terms (Objectives 8 - 14)
Anatomical position (Objective 8)
- The body is erect with feet flat on the floor and facing forward; arms are at the sides; palms and toes are directed forward.

Directions in the body (Objective 9)
- Six pair of opposite terms are used to describe the relative position of one body part to another: superior/inferior; anterior/posterior; medial/lateral; proximal/distal; superficial/deep; visceral/parietal.

Planes and sections of the body (Objective 10)
- A sagittal plane divides the body into right and left parts.

- A transverse, or horizontal, plane divides it into upper and lower regions.
- A frontal, or coronal, plane divides it into front and back portions.

Body cavities (Objectives 11 and 12)
- The dorsal body cavity consists of the cranial cavity, which contains the brain, and the spinal cavity, which contains the spinal cord.
- The ventral body cavity is subdivided into the thoracic cavity, with contains the heart and lungs, and the abdominopelvic cavity, which contains the digestive, urinary, and reproductive organs.
- A convenient and commonly used method divides the abdominopelvic cavity into nine regions: epigastric, umbilical, hypogastric, right and left hypochondriac, right and left lumbar, and right and left iliac.

Regions of the body (Objectives 13 and 14)
- The axial portion of the body consists of the head, neck, and trunk.
- The appendicular portion consists of the limbs or appendages.
- Specific anatomic terms are used to designate body regions.

☞ **Answers to Review Questions**

1. Anatomy and physiology are interrelated because structure influences function and function affects structure.
2. Chemical, cellular, tissue, organ, body system, and total organism.
3. **Integumentary:** Covers and protects the body, regulates temperature.
   Skeletal: Provides body framework and support, protects, attaches muscles to bones, calcium storage.
   Muscular: Produces movement, maintains posture, provides heat.
   Nervous: Coordinates body activities, receives and transmits stimuli.
   Endocrine: Regulates metabolic activities and body chemistry.
   Cardiovascular: Transports material from one part of the body to another, defends against disease.
   Lymphatic: Returns tissue fluid to the blood and defends against disease.
   Digestive: Ingests and digests food, absorbs nutrients into blood.
   Respiratory: Exchanges gases between blood and external environment.
   Urinary: Excretes metabolic wastes, regulates fluid balance and regulates acid-base balance.

Reproductive: Forms new individuals to provide continuation of the human species.

4. Organization, metabolism, responsiveness, movement, and reproduction.

5. Growth, differentiation, respiration, digestion, and excretion.

6. Water, oxygen, nutrients, heat, and pressure.

7. Homeostasis means "staying the same." It refers to maintaining a consistent internal environment. Negative feedback mechanisms maintain homeostasis by keeping variations within a normal range.

8. Positive feedback mechanisms differ from negative feedback mechanisms in that the positive feedback mechanisms are stimulatory to bring a process to a rapid conclusion.

9. The body is in anatomical position when it is erect with head facing forward, arms are at the sides with palms of the hands facing forward, feet parallel with toes facing forward.

10. The **mouth** is **inferior** to the nose; or the **nose** is **superior** to the mouth.
    The **spinal cord** is **posterior** to the heart; or the **heart** is **anterior** to the spinal cord.
    The **ears** are **lateral** to the eyes; or the **eyes** are **medial** to the ears.
    The **elbow** is **proximal** to the wrist; or the **wrist** is **distal** to the elbow.
    The **skin** is **superficial** to the muscles; or the **muscles** are **deep** to the skin.

11. Coronal plane divides into anterior and posterior portions.
    Transverse plane divides into superior and inferior portions.
    Sagittal plane divides into right and left portions.

12. The dorsal body cavity and the ventral body cavity; the ventral body cavity is larger.

13.

A	B	C
D	E	F
G	H	I

A. Right hypochondriac region
B. Epigastric region
C. Left hypochondriac region
D. Right lumbar region
E. Umbilical region
F. Left lumbar region
G. Right iliac (inguinal) region
H. Hypogastric region
I. Left iliac (inguinal) region

14. Head, neck, and trunk

15. Head
    Cranial refers to the region that encloses the brain.
    Ophthalmic refers to the eyes.
    Buccal refers to the cheeks.
    Frontal refers to the forehead.
    Otic refers to the ears.
    Oral refers to the mouth.
    Arm
    Brachial refers to the region between the shoulder and elbow.
    Antecubital refers to the region anterior to the elbow.
    Thorax
    Sternal refers to the anterior midline of the thorax.
    Mammary refers to the breast.
    Costal refers to the ribs.
    Abdomen
    Celiac refers to the abdominal region
    Umbilical refers to the middle region of the abdomen.
    Navel is the same as the umbilical region.
    The inguinal region, or groin, is the depressed region between the abdomen and thigh.
    Lower extremity
    The thigh. or femoral region is the area between the hip and knee.
    The leg, or crural, region is the are between the knee and ankle.
    The popliteal region is the area behind the knee.
    Tarsal refers to the ankle and instep of the foot.
    Pedal refers to the foot.
    Plantar refers to the sole of the foot.

☞ **Answers to Learning Exercises**

Anatomy and Physiology (Objective 1)

1. Anatomy
2. Physiology
3. Function, structure
4. Surface anatomy
   Cytology
   Embryology
   Immunology
   Pharmacology
   Pathology

Levels of Organization (Objective 2)

1. Chemical, cellular, tissues, organs, body systems, total organism.
2. Cell
3. Tissue

Organ Systems (Objective 3)
1. Integumentary
2. Skeletal
3. Digestive
4. Respiratory
5. Urinary
6. Endocrine
7. Lymphatic
8. Nervous
9. Cardiovascular
10. Lymphatic
11. Digestive
12. Nervous
13. Endocrine
14. Cardiovascular
15. Urinary
16. Muscular
17. Lymphatic
18. Integumentary
19. Skeletal
20. Reproductive

Life Processes (Objectives 4 and 5)
1. (Any order) Organization, metabolism, responsiveness, movement, reproduction, growth, differentiation, respiration, digestion, excretion.
2. Catabolism
3. Anabolism is a building up process in which complex substances are synthesized from simpler ones.
4. (Any order) Water, oxygen, nutrients, heat, pressure.

Homeostasis (Objectives 6 and 7)
1. Homeostasis
2. Stressor
3. Negative feedback
4. Positive feedback

Anatomical Terms (Objectives 8-14)
1. Body is **erect**.
   Arms are **at sides**.
   Feet and toes are **directed forward**.
   Face is **forward**.
   Palms are **forward**.
2. Superior
   Proximal
   Deep
   Anterior to
   Visceral
3. Midsagittal (median sagittal)
   Transverse (horizontal)
   Frontal (coronal)
   Sagittal
4. Dorsal cavity
   Ventral cavity
   Cranial cavity
   Thoracic cavity
   Abdominal cavity
   Spinal cavity
   Pelvic cavity
5. A. Right hypochondriac
   B. Umbilical
   C. Left iliac (inguinal)
   D. Epigastric
   E. Left lumbar
   F. Left hypochondriac
   G. Right lumbar
   H. Hypogastric
   I. Right iliac (inguinal)
6. A. Otic
   B. Sternal
   C. Brachial
   D. Antebrachial
   E. Leg (crural)
   F. Oral
   G. Pectoral (mammary)
   H. Umbilical (navel)
   I. Inguinal
   J. Femoral
   K. Occipital
   L. Sacral
   M. Carpal
   N. Popliteal
   O. Pedal
   P. Cervical
   Q. Axillary
   R. Lumbar
   S. Gluteal
   T. Palmar

☞ Answers to Chapter Self-Quiz
1. Physiology
2. (b) cell
3. (d) lymphatic/heart/defense against disease
4. Catabolism
5. (d) oxygen, water, pressure
6. (e) increases deviations from normal
7. (b) your eyes are facing the same direction as your palms
8. (c) proximal
9. Frontal (coronal)
10. (d) epigastric
11. Heart, thoracic cavity, small intestines should have an X.
12. (b) arms and legs
13. D Skull
    E Buttock region
    B Armpit area
    G Chest region
    H Area behind the knee
    J Middle region of abdomen

I   Anterior midline of thorax
A   Space in front of the elbow
F   Mouth
C   Neck region

☞ Answers to Terminology Exercises

Dorsal
Gastritis
Cardiology
Homeostasis
Metabolism

Nearest to
Pertaining to internal organs
Above the stomach
Pertaining to a dried, hard body
Study of disease

D (process of cutting apart)
E (skin or covering of body)
A (study of function)
B (structure shaped like a basin)
C (pertaining to heart/blood vessels)

☞ Answers to Fun and Games

```
L A M I X O R P • • • • • • • • • •
A • • M • • • • • C • Y • • R L • • • F
E R • E • • • • A A R L S A O A • C N R
T O • T • • • T N O A A • B I P • E • O
I I • A • • A A T T G C • D R R G L • N
L R • B • B B A E I • C • O E A • L • T
P E • O O O R R T S • U • M T C • • • A
O P • L L I A T • S • B V I S C E R A L
P U I I P L A • • U • • V N O • • • • •
• S S S • L • N • E • E L O P • Y • U •
M M E M • • F Y T • F A • P • G • M • •
Y R A T N E M U G E T N I E O • B • E R
• • • • M O • C E I B • • L • I S • V O
• • • O T • I D P • • R O V L • K • I S
• • • R A • C B I • • I A I • • E • T S
• A N • A A C • • • S • C C • • L • S E
L A • R C C • • • Y • A • • H • E • E R
• • O K O • • • H • L • • • • I T • G T
• H • • • • • P • H O M E O S T A S I S
T • • • • • • • • • • • • • • L L D •
```

Word List

1. Abdominopelvic
2. Anabolism
3. Anatomy
4. Antebrachial
5. Buccal
6. Carpal
7. Catabolism
8. Cell
9. Digestive
10. Femoral
11. Frontal
12. Homeostasis

13. Integumentary
14. Lateral
15. Metabolism
16. Negative feedback
17. Occipital
18. Physiology
19. Popliteal
20. Posterior
21. Proximal
22. Respiratory
23. Sagittal
24. Skeletal
25. Stressor
26. Superior
27. Thoracic
28. Tissue
29. Umbilical
30. Visceral

☞ Quiz/Test Questions

Note: There are fifty multiple-choice questions for this chapter in the computerized test bank.

Name the following:

1. Plane that divides an organ into right and left portions.
   Answer: sagittal.

2. Term that means closer to origin or attachment.
   Answer: proximal.

3. Term for the neck region.
   Answer: cervical.

4. Term for the region behind the knee.
   Answer: popliteal.

5. Term for the arm.
   Answer: brachium.

6. Three physical factors necessary for human life.
   Answer: (any three) water, nutrients, oxygen, heat, and pressure.

7. The body system that includes the skin.
   Answer: integumentary.

8. Organizational level that is a collection of different tissues that work together to perform one or more functions.
   Answer: organ.

9. Type of mechanism in which actions stimulate each other until there is a culminating event that terminates the process.
   Answer: positive feedback.

10. Study of functions and their relationships to each other and to structure.
    Answer: physiology.

Questions for this chapter continue on page 135.

2 Chemistry, Matter, and Life

Acid A substance that ionizes in water to released hydrogen ions; a proton donor; a substance with a pH less than 7.0.

Anion A negatively charged ion.

Atom The smallest unit of a chemical element that retains the properties of that element.

Atomic number The number of protons in the nucleus of an atom of an element.

Base A substance that ionizes in water to release hydroxyl (OH⁻) ions or other ions that combine with hydrogen ions; a proton acceptor; a substance with a pH greater than 7.0 (alkaline).

Buffer A substance that reduces the change in pH when either an acid o a base is added.

Carbohydrate An organic compound that contains carbon, hydrogen, and oxygen with the hydrogen and oxygen present in a 2:1 ratio; sugar, starch, cellulose.

Catalyst A substance that speeds up chemical reactions without being changed itself.

Cation A positively charged ion.

Compound A substance formed from two or more elements joined by chemical bonds in a definite, or fixed, ratio; smallest unit of a compound is a molecule.

Covalent bond Chemical bond formed by two atoms sharing one or more pairs of electrons.

Electrolyte A substance that forms positive and negative ions in a solution, which makes it capable of conducting an electric current.

Electron A negatively charged particle found in the nucleus of an atom.

Element Simplest form of matter that cannot be broken down by ordinary chemical means.

Ion Electrically charged atom or group of atoms; an atom that has gained or lost one or more electrons.

Ionic bond Chemical bond that is formed when one or more electrons are transferred from one atom to another.

Lipid A class of organic compounds that includes oils, fats, and related substances.

Mass number The total number of protons and neutrons in the nucleus of an atom of an element.

Matter Anything that has weight and takes up space.

Molecule A particle composed of two or more atoms that are chemically bound together; smallest unit of a compound.

Neutron An electrically neutral particle found in the nucleus of an atom.

Protein An organic compound that contains nitrogen and consists of chains of amino acids linked together by peptide bonds.

Proton A positively charged particle found in the nucleus of an atom.

Solute A substance that is dissolved in a solution.

Solvent Fluid in which substances dissolve.

☞ Chapter Objectives

Upon completion of this chapter the student should be able to:

1. Define matter.
2. Define an element.
3. Use chemical symbols to identify elements.
4. Differentiate between protons, neutrons, and electrons, and tell where each one is located.
5. Draw a simplified diagram that illustrates the structure of an atom.
6. Distinguish between atomic number and mass number of an element.
7. Describe the electron arrangement that makes an atom most stable.
8. Describe the difference between ionic bonds, covalent bonds, and hydrogen bonds.
9. Distinguish between cations and anions.
10. Describe the relationship between atoms, molecules, and compounds.
11. Interpret molecular formulas for compounds.
12. Identify the reactants and products in a chemical equation.
13. Describe and illustrate four types of chemical reactions.
14. Compare exergonic and endergonic reactions.
15. Discuss five factors that influence the rate of chemical reactions.
16. Explain what is meant by a reversible reaction.
17. Distinguish between mixtures, solutions, and suspensions.

18. Define the term "electrolyte."
19. Describe what makes an acid or a base and what happens when they react.
20. Discuss the concepts of pH and buffers.
21. Describe the five major groups of organic compounds that are important to the human body.

☞ **Chapter Outline/Summary**

Elements (Objectives 1 - 3)
- Matter is anything that takes up space and has weight.
- The simplest form of matter is an element.
- Chemical symbols are abbreviations used to identify elements.

Structure of Atoms (Objectives 4 - 7)
- An atom is the smallest unit of an element.
- The atomic number of an atom is the number of positively charged particles, protons, in the nucleus.
- Neutrons, also in the nucleus, have the same mass as protons, but have no charge. The mass number of an atom equals the number of protons plus the number of neutrons.
- Electrons are negatively charged particles that are in constant motion outside the nucleus. The most stable atoms have eight electrons in their highest energy level.

Chemical Bonds (Objectives 8 and 9)
- Chemical bonds are forces that hold atoms together

Ionic bonds
- An ion is an atom that has lost or gained one or more electrons.
- A cation is a positively charged ion.
- An anion is a negatively charged ion.
- Ionic bonds are the attraction forces between cations and anions to form ionic compounds.

Covalent bonds
- Covalent bonds result when atoms share electrons.
- Atoms may share more than one pair of electrons, which results in double or triple covalent bonds.
- An unequal sharing of electrons results in polar covalent bonds.

Hydrogen bonds
- Hydrogen bonds are intermolecular bonds that are attractions between molecules
- Hydrogen bonds are formed by the attraction between the electropositive hydrogen end of a polar covalent compound and the negative charges on other molecules or ions.

Compounds and Molecules (Objectives 10 - 11)
Nature of compounds (Objective 10)
- The atoms in a molecule are held together by chemical bonds.
- Atoms combine in definite ratios to form molecules, which are the smallest units of compounds.

Formulas (Objective 11)
- Molecular formulas use chemical symbols to indicate the type of atoms and numerical subscripts to show how many of each atom are in a molecule of a compound.

Chemical Reactions (Objectives 12 - 16)
Equations (Objective 12)
- Chemical equations are an abbreviated method of showing the reactants and products in a chemical reaction

Types of reactions (Objectives 13 and 14)
- Synthesis reactions form a complex molecule from two or more simple molecules.
- Decomposition reactions break down large molecules into simpler ones.
- In single and double replacement reactions, the reactants exchange one or more elements to form new compounds.
- Exergonic reactions release energy. Endergonic reactions require energy which is then stored in the chemical bonds.

Rate of reactions (Objective 15)
- The nature of the reacting substances affects the reaction rate.
- Reaction rates increase as temperature increases.
- Increasing the concentration of the reactants to an optimum increases the rate of the reaction.
- Enzymes and other catalysts increase the rate of a reaction.
- Breaking the reactants into small particles increases the total surface area of the particles and increases the reaction rate.

Reversible reactions (Objective 16)
- Some chemical reactions are reversible and the direction they proceed depends on the existing conditions.

Mixtures, Solutions, and Suspensions (Objective 17)
- A mixture consists of two or more substances that can be physically separated.
- Solutions consist of a solute that is being dissolved and a solvent that does the dissolving.
- In most suspensions, the particles settle if left undisturbed. In colloidal suspensions, the particles are so small they remain suspended but do not dissolve.

Acids, Bases, and Buffers (Objectives 18, 19, and 20)
- Electrolytes form positive and negative ions when they are dissolved in water.
- Acids are proton donors.
- Bases accept protons.
- pH values indicate the hydrogen ion concentration of a solution. A pH of 7 is neutral. Acids have a pH less than 7, bases have a pH greater than 7.
- Neutralization reactions occur between acids and bases to produce salts and water.
- Buffers, which contain a weak acid and a salt of that same acid, resist pH changes by neutralizing the effects of stronger acids and bases.

Organic Compounds (Objective 21)
Carbohydrates
- Carbohydrates contain carbon, hydrogen, and oxygen. They are an important energy source.
- Glucose, fructose, and galactose are monosaccharides, or simple sugars, with 6 carbon atoms. Ribose and deoxyribose have 5 carbon atoms.
- Sucrose, maltose, and lactose are disaccharides or double sugars. They consist of two hexose monosaccharides linked together.
- Starch, cellulose, and glycogen are important polysaccharides. They consist of long chains of glucose molecules.

Proteins
- Proteins are formed from amino acids linked together by peptide bonds. They contain carbon, hydrogen, oxygen, nitrogen, usually sulfur, and often phosphorus.
- Proteins are important to the health of an individual.

Lipids
- Lipids are insoluble in water but will dissolve in solvents such as alcohol and ether. They contain carbon, hydrogen, and oxygen.
- The building blocks of triglycerides, commonly known as fats, are glycerol and fatty acids. Saturated fats contain only fatty acids that have single bonds between the carbon atoms.
- Phospholipids, which contain phosphates and nitrogen, are important components of cell membranes.
- Steroids, which are derivatives of lipids, include cholesterol, certain hormones, and vitamin D.

Nucleic acids
- Nucleotides are the building blocks of nucleic acids. They contain carbon, hydrogen, oxygen, nitrogen, and phosphorus.
- DNA is the genetic material of the cell and RNA functions in the synthesis of proteins within the cell.

Adenosine triphosphate
- ATP is a high energy compound that supplies energy in a form that is usable by body cells.

☞ **Answers to Review Questions**

1. Matter is anything that has mass and takes up space.
2. An element is the simplest form of matter.
3. (a) O; (b) N; (c) Ca; (d) K; (e) Na; (f) Fe
4. a. neutron
   b. electron
   c. electron
   d. proton
   e. proton and neutron
5.

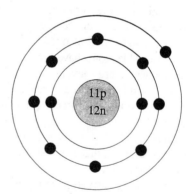

6. a. 35
   b. 17
7. 8
8. An ionic bond is formed when electrons are lost or gained. A covalent bond is formed when atoms share electrons.
9. Cations are atoms that have lost one or more electrons and have a positive charge; anions are atoms that have gained one or more electrons and have a negative charge.
10. A compound consists of two or more different types of atoms chemically combined in a definite ratio. A molecule is the smallest unit of a compound.
11. a. Sodium (Na), hydrogen (H), carbon (C), and oxygen (O).
    b. One atom each of sodium, hydrogen, and carbon. Three atoms of oxygen.

12. a. Reactants are $C_6H_{12}O_6$ and O_2.
    b. Products are CO_2 and H_2O.
13. a. Synthesis, combination, or composition.
    b. Single replacement (displacement).
    c. Decomposition.
    d. Double replacement (displacement) or exchange.
14. Exergonic
15. Increase the temperature; add a catalyst; increase the concentration of the reactants; break the reactants into smaller pieces to increase surface area.
16. Exhaling CO_2 drives the reaction to the left and reduces the number of hydrogen ions in the body.
17. The components of a mixture are combined in varying proportions and can be separated by physical means. The components of a compound are combined in definite ratios and cannot be separated by physical means.
18. A solution is clear and the solute does not settle. A suspension is cloudy and its particles settle.
19. An electrolyte is a substance that breaks up or dissociates in water to form charged particles (ions) that conduct electricity. The electrocardiogram (ECG), electroencephalogram (EEG), and electromyogram (EMG) all depend on electrolyte activity.
20. An acid is a proton (hydrogen ion) donor. A base is a proton (hydrogen ion) acceptor.
21. Reactants: acids and bases
    Products: water and salts
22. Acids have pH < 7
    Bases have pH > 7
23. A buffer solution contains a weak acid and a salt of that same acid. The weak acid reacts with strong bases to make a weak base (salt) and water. The salt reacts with strong acids to form a weaker acid and a neutral salt.
24. Carbohydrates, proteins, lipids, nucleic acids, and ATP.
25. Monosaccharides:
    Glucose--sugar found in blood
    Fructose--fruit sugar; sweetest sugar
    Galactose--found as part of lactose in milk
    Disaccharides:
    Sucrose--table sugar; glucose + fructose
    Maltose--malt sugar; 2 glucose
    Lactose--milk sugar; glucose + galactose
    Polysaccharides:
    Starch--storage form of glucose in plants
    Glycogen--storage form of glucose in animals
    Cellulose--forms supporting fiber in plants; not digestible

26. Proteins are long chains of amino acids linked together by peptide bonds. They contain carbon, hydrogen, oxygen, and nitrogen. Many also contain sulfur and/or phosphorus
27. A triglyceride is glycerol with three fatty acids attached, one at each -OH group. A saturated fat (triglyceride) contains only saturated fatty acids. An unsaturated fat (triglyceride) contains one or more unsaturated fatty acids.
28. Phospholipids, an important component of cell membranes, contain phosphorus and nitrogen in addition to the fatty acids and glycerol. Steroids are derivatives of lipids that contain four interconnected carbon rings. Cholesterol is the most common steroid.
29. DNA: Sugar is deoxyribose, bases are thymine, cytosine, guanine, and adenine, the molecule is a double strand. RNA: Sugar is ribose, uracil replaces thymine as a base, and the molecule is a single strand.

☞ **Answers to Learning Exercises**

Elements (Objectives 1-3)
1. Matter is defined as anything that takes up space and has mass.
2. An element is defined as the simplest form of matter; cannot be broken into simpler form by ordinary chemical means.
3. H = hydrogen; K = Potassium; O = Oxygen; C = Carbon; Mg = Magnesium; P = Phosphorus
4. Sodium = Na; Chlorine = Cl; Iron = Fe; Calcium = Ca; Nitrogen = N; Sulfur = S

Structure of Atoms (Objectives 4-7)
1.

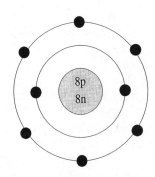

2. Particle: neutron, electron, proton
   Location: nucleus, orbitals, nucleus
   Charge: 0, −. +
   Mass: 1, negligible, 1

3. 19 protons, 20 neutrons, 19 electrons

4. 8 electrons

Chemical Bonds (Objectives 8-9)

1. C, D, E, A, B, C

Compounds and Molecules (Objectives 10-11)

1. Atom
   Molecule
   Compound
   Molecule

2. Calcium, 1 atom
   Carbon, 1 atom
   Oxygen, 3 atoms

Chemical Reactions (Objectives 12-16)

1. Equation: $H_2CO_3 \rightarrow H_2O + CO_2$
   Type of reaction: Decomposition
   Reactants: H_2CO_3
   Products: H_2O and CO_2

   Equation: $N_2 + 3H_2 \rightarrow 2NH_3$
   Type of reaction: Composition, Synthesis
   Reactants: N_2 and H_2
   Products: NH_3, ammonia

   Equation:
   $MgCl_2 + 2NaOH \rightarrow Mg(OH)_2 + 2NaCl$
   Type of reaction: Double replacement
   Reactants: $MgCl_2$ and NaOH
   Products: $Mg(OH)_2$ and NaCl

   Equation:
   $C_7H_6O_3 + C_2H_4O_2 \rightarrow C_9H_8O_4 + H_2O$
   Type of reaction: Composition, Synthesis
   Reactants: $C_7H_6O_3$ and $C_2H_4O_2$
   Products: $C_9H_8O_4$ and H_2O

2. Exergonic, endergonic

3. I (Grind up the reactants)
   D (Dilute one reactant)
   D (Cool the reaction mixture)
   I (Increase the temperature)
   I (Use more concentrated solutions)
   I (Increase the temperature)

4. The reaction is reversible. It may proceed in either direction

Mixtures, Solutions, and Suspension (Objective 17)

1. (c) solvent

2. (a) are clear

3. Mixture (sugar and salt)
   Solution (sugar and water)
   Mixture (sand and water)
   Suspension (blood cells and plasma)
   Suspension (cytoplasm of the cell)

Acids, Bases, and Buffers (Objectives 18-20)

1. Electrolytes

2. Base (accepts hydrogen ions)
   Acid (has sour taste)
   Acid (has pH of 3.5)

Acid (donates protons)
Acid (reacts with a buffer to form weak acid)
Acid (reacts with OH^- ions to form water)
Base (has slippery, soapy feeling)
Base (has pH of 8.7)

3. $HC_2H_3O_2$ reacts; H_2O is formed
   $NaC_2H_3O_2$ reacts; NaCl is formed

Organic Compounds (Objective 21)

1. Carbon, hydrogen, oxygen

2. Glucose, fructose, and galactose

3. Disaccharide

4. Complex polysaccharides

5. Nitrogen
   Amino acids
   Essential

6. Glycerol and fatty acids
   Saturated
   Phospholipids

7. DNA
   RNA
   Nucleotides
   Deoxyribose
   DNA
   C, H, O, N, P
   RNA
   ATP

8. A (glucose)
   B (steroids)
   B (glycerol)
   C (hemoglobin)
   A (disaccharides)
   C (amino acids)
   D (nucleotides)
   A (glycogen)
   D (RNA)
   B (triglycerides)

☞ **Answers to Chapter Self-Quiz**

1. (a) Carbon
   (b) Phosphorus
   (c) Chlorine
   (d) Iron
   (e) Copper
   (f) Oxygen

2. (a) Atomic number = 17
   (b) Neutrons = 18
   (c) Protons = 17

3. C
   B
   E
   A
   D

4. (a) Zinc, copper, sulfur, oxygen

(b) $Zn + CuSO_4$
(c) 4 atoms of oxygen
(d) Single replacement (displacement)

5. A (proton donor)
   B (pH = 8.2)
   B (accepts hydrogen ions)
   A (pH = 2.5)
   B (accepts protons)
   A (pH = 6.5)
   A (donates hydrogen ions)

6. Acid is HCl
   Base is NaOH
   Salt is NaCl

7. F (remove some product)
   F (add catalyst)
   S (decrease temperature)
   F (break up reactants)
   F (increase concentration)

8. P (cellulose)
   M (fructose)
   M (galactose)
   M (glucose)
   P (glycogen)
   D (lactose)
   D (maltose)
   D (sucrose)

9. · N (adenine, cytosine)
   · P (amino acids)
   , L (cholesterol)
   C (monosaccharides)
   N (nucleotides)
   P (peptide bonds)
   C (starch)
   L (triglycerides)

10. e

☞ **Answers to Terminology Exercises**

Polysaccharide
Lipogenesis
Diphosphate
Exergonic
Dioxide

Molecule with 3 phosphates
Putting energy into
Two sugars
Milk sugar
To take apart with water

D (forming glycogen)
C (less oxygen than ribose)
A (basic, pH > 7)
B (pertaining to carbon and water)
E (breaking down fat)

☞ **Answers to Fun and Games**

A. Molecule
   Proton
   Amino acid
   Base
   Salt
   Catalyst
   Hydrogen
   Polysaccharide
   Final Word: Adenosine triphosphate

B. Oxygen
   Ribose
   Nitrogen
   Lipids
   Adenine
   Cation
   Electron
   Buffer
   Triglyceride
   Final Word: Deoxyribonucleic acid

☞ **Quiz/Test Questions**

Note: There are fifty multiple-choice questions for this chapter in the computerized test bank.

Name the following:

1. Simplest form of matter than cannot be broken down by ordinary chemical means.
   Answer: element.

2. Element that is designated by the symbol K.
   Answer: potassium.

3. Negatively charged particle in an atom.
   Answer: electron.

4. Mass number of an atom with 8 protons, 8 neutrons, and 8 electrons.
   Answer: 16.

5. Chemical bond that is formed when electrons are shared.
   Answer: covalent.

6. Positively charged ions.
   Answer: cations.

7. Elements and total number of atoms in $NaHCO_3$.
   Answer: elements are sodium, hydrogen, carbon, and oxygen; total number = 6.

8. Compound that is a product in this reaction: $Cl_2 + 2NaBr \rightarrow 2NaCl + Br_2$.
   Answer: NaCl; remember Br_2 is not a compound because it has only one type of atom.

Questions for this chapter continue on page 135.

3 Cell Structure and Function

☞ Key Terms/Concepts

Active transport Membrane transport process that requires cellular energy (ATP).

Anticodon A sequence of three nucleotide bases on transfer RNA that is complementary to a codon on messenger RNA; represents a single amino acid.

Cell membrane Phospholipid membrane that separates the contents of the cell from the material outside the cell.

Centrosome Dense area near the nucleus that contains the centrioles.

Chromatin Long, slender threads of DNA in the nucleus of a cell; gives rise to chromosomes during mitosis.

Chromosomes Dark staining structures that appear in the nucleus when chromatin condenses during mitosis.

Codon A set of three nucleotides on a messenger RNA molecule; represents a single amino acid.

Cytokinesis Division of the cytoplasm at the end of mitosis to form two separate daughter cells.

Cytoplasm Gel like fluid inside the cell, exclusive of the organelles.

Diffusion Movement of atoms, ions, or molecules from a region of high concentration to a region of low concentration.

Endoplasmic reticulum Membrane enclosed channels within the cytoplasm.

Filtration The movement of a fluid through a membrane in response to hydrostatic pressure.

Golgi apparatus Membranous sacs within the cytoplasm that process and package cellular products.

Lysosome Membrane enclosed sac of digestive enzymes within the cytoplasm.

Meiosis Type of nuclear division in which the number of chromosomes is reduced to one half the number found in a body cell; results in the formation of an egg or sperm.

Mitochondria Organelles that contain the enzymes essential for producing ATP; singular, mitochondrion.

Mitosis Process by which the nucleus of a body cell divides to form two new cells, each identical to the parent cell.

Nucleolus A dense, dark staining body within the nucleus; contains a high concentration of RNA.

Nucleus Largest structure within the cell; contains the DNA.

Organelles Little organs; highly organized structures suspended in the cytoplasm that are specialized to perform specific cellular activities.

Osmosis Diffusion of water through a selectively permeable membrane.

Passive transport Membrane transport process that does not require cellular energy.

Phagocytosis Cell eating; a form of endocytosis in which solid particles are taken into the cell.

Pinocytosis Cell drinking; a form of endocytosis in which fluid droplets are taken into the cell.

Ribosome Granules of RNA in the cytoplasm that function in protein synthesis.

☞ Chapter Objectives

Upon completion of this chapter the student should be able to:

1. Explain what is meant by a "generalized cell."
2. Describe the composition of the cell membrane.
3. List five functions of the proteins in the cell membrane.
4. Describe the cytoplasm.
5. Describe the components of the nucleus and state the function of each one.
6. Describe each of the cytoplasmic organelles and state the function of each one.
7. Characterize the cytoskeleton.
8. Relate the structure, location, and function of the centrioles.
9. Distinguish between cilia and flagella on the basis of structure and function.
10. Identify the parts of a generalized cell.
11. Explain how the cell membrane regulates the composition of the cytoplasm.
12. Describe the process of diffusion and give a physiologic example.

13. Distinguish between simple diffusion and facilitated diffusion.
14. Explain the difference between simple diffusion and osmosis.
15. Explain the significance of isotonic, hypertonic, and hypotonic solutions.
16. Describe the mechanics of filtration and cite two physiologic examples.
17. Distinguish between active transport and other types of membrane transport.
18. Distinguish between endocytosis and exocytosis and give examples.
19. Name the phases of a typical cell cycle and describe the events that occur in each phase.
20. Explain the difference between mitosis and meiosis.
21. Explain what constitutes a gene.
22. Describe the process of DNA replication.
23. Explain how DNA in the nucleus regulates protein synthesis in the cytoplasm by relating the events that occur during protein synthesis.
24. Define the terms transcription and translation as they pertain to protein synthesis.

☞ **Chapter Outline/Summary**

Structure of the Generalized Cell
(Objectives 1 -10)
- There are many different types, sizes, and shapes of cells in the body. For descriptive purposes, the concept of a "generalized cell" is introduced. It includes features from all cell types.

Cell membrane (Objectives 2 and 3)
- A selectively permeable cell membrane separates the extracellular material from the intracellular material.
- The cell membrane is a double layer of phospholipid molecules.
- Proteins in the cell membrane provide structural support, form channels for passage of materials, act as receptor sites, function as carrier molecules, and provide identification markers.

Cytoplasm (Objective 4)
- Cytoplasm, the gel-like fluid inside the cell, is largely water and has a variety of organelles suspended in it.

Nucleus (Objective 5)
- The nucleus, formed by a nuclear membrane around a fluid nucleoplasm, is the control center of the cell.
- Threads of chromatin in the nucleus contain DNA, the genetic material of the cell.

- The nucleolus is a dense region of RNA in the nucleus and is the site of ribosome formation.

Cytoplasmic organelles (Objective 6)
- Mitochondria are enclosed by a double membrane and function in the production of ATP.
- Ribosomes are granules of RNA that function in protein synthesis.
- Endoplasmic reticulum is a series of membranous channels that function in the transport of molecules. Rough endoplasmic reticulum has ribosomes associated with it and it transports proteins. Smooth endoplasmic reticulum doesn't have ribosomes and it transports certain lipids.
- Golgi apparatus modifies substances that are produced in other parts of the cell and prepares these products for secretion.
- Lysosomes contain enzymes that break down substances taken in at the cell membrane. They also destroy cellular debris.

Filamentous protein organelles (Objectives 7-9)
- Cytoskeleton is formed from microfilaments and microtubules and helps to maintain the shape of the cell.
- Centrioles are paired and are located in the centrosome, a dense region near the nucleus. Centrioles function in cell division.
- Cilia are short, hair-like projections that move substances across the surface of a cell.
- Flagella are long, thread-like, projections that move the cell.

Cell Functions (Objectives 11 - 24)
- The functions of specific cells are closely related to the structure of those cells.

Movement of substances across the cell membrane (Objectives 11 - 18)
- The cell membrane controls the composition of the cytoplasm by regulating movement of substances through the membrane.
- Diffusion is the movement of particles from a region of higher concentration to a region of lower concentration. It may take place through a permeable membrane, but also occurs when no membrane is involved.
- Facilitated diffusion requires a special carrier molecule but still goes from a region of higher concentration on one side of the membrane to a region of lower concentration on the other side.
- Osmosis is the diffusion of solvent or water molecules through a selectively permeable membrane. Cells placed in a hypotonic solution will take in water by osmosis and

will swell due to the increased intracellular volume. Cells placed in a hypertonic solution will lose fluid due to osmosis and will shrink or crenate.

- Filtration utilizes pressure to push substances through a membrane. The pores in the membrane filter determine the size of particles that will pass through it.
- Active transport moves substances against a concentration gradient, from a region of lower concentration to a region of higher concentration. It requires a carrier molecule and uses energy.
- Endocytosis occurs when solid particles and liquid droplets are taken into the cell. Phagocytosis (solids) and pinocytosis (liquids) are types of endocytosis.
- Exocytosis is the process by which secretory vesicles are moved from the inside to the outside of the cell.

Cell division (Objectives 19 and 20)

- New cells are continually needed for growth, repair, and replacement. There are two types of cell division. Somatic cells reproduce by mitosis, which results in two cells identical to the one parent cell.
- Interphase is the period between successive mitotic cell divisions. It is the longest part of the cell cycle.
- The successive stages of mitosis are prophase, metaphase, anaphase, and telophase. Cytokinesis, division of the cytoplasm, occurs during telophase.
- Reproductive cells divide by meiosis. In meiosis, a single parent cell produces four cells, each with one-half the number of chromosomes as the parent cell.

DNA replication and protein synthesis (Objectives 21 - 24)

- DNA in the nucleus directs protein synthesis in the cytoplasm. A gene is the portion of a DNA molecule that controls the synthesis of one specific protein molecule.
- When DNA replicates, the hydrogen bonds between the complimentary base pairs break, resulting in two single strands. The specific pairing of nitrogenous bases results in the formation of new complimentary strands. When replication is completed, there are two double strand DNA molecules, each one identical to the original.
- In protein synthesis messenger RNA carries the genetic information from the DNA in the nucleus to the sites of protein synthesis in the cytoplasm. During the process of transcription, the genetic code is transferred from DNA to mRNA.
- A sequence of three nucleotide bases on mRNA represents a codon that codes for a specific amino acid. Anticodons on tRNA have nitrogenous bases that are complimentary to the codons on mRNA.
- Each tRNA carries a specific amino acid to the developing molecule. The amino acid sequence of the developing protein molecule is determined by the complimentary base pairs on the mRNA and tRNA. The process of creating a protein in response to mRNA codons is called translation.

☞ **Answers to Review Questions**

1. A cell, used for descriptive purposes, that contains the components of many different cell types.
2. Phospholipids and proteins; the phospholipids are arranged as a bilayer with the phosphates forming the two surfaces and the lipids sandwiched between these surfaces; the proteins are scattered among the phospholipids.
3. The proteins provide structural support, form channels for passage of materials, act as receptor sites, function as carrier molecules, and provide identification markers (antigens).
4. Cytoplasm
5. Chromatin and the nucleolus are located inside the nucleus of the cell. Chromatin is long filamentous strands of DNA, the genetic material of the cell. During cell replication, chromatin becomes chromosomes. The nucleolus is a dark-staining body within the nucleus. It contains RNA and is where ribosomes are formed.
6. **Mitochondria** are oval fluid-filled sacs enclosed by a double membrane and function in the production of ATP. The inner membrane invaginates to form cristae.
   Ribosomes are granules of RNA that function in protein synthesis.
   Rough endoplasmic reticulum is a series of membranous channels with ribosomes attached. It functions in protein transport.
   Smooth endoplasmic reticulum is similar to rough endoplasmic reticulum except there are no ribosomes attached. It functions in the synthesis and transport of certain lipids such as steroids.

Golgi apparatus is a series of 4-6 flattened membranous sacs near the nucleus. It prepares cellular products for secretion.

Lysosomes are membrane-enclosed sacs of digestive enzymes. They destroy cellular debris and harmful substances that are taken into the cell.

7. Microfilaments and microtubules

8. Centrioles are located within the centrosome near the nucleus. They function in the distribution of chromosomes to the daughter cells in cell division.

9. Cilia are short, numerous, and move substances across the surface of the cell. Flagella are long, usually occur singly, and function to move the cell itself.

10. Right hand column in sequence:
    Cilia
    Rough endoplasmic reticulum
    Ribosomes
    Lysosome
    Chromatin
    Microtubules
    Nucleus
    Golgi apparatus
    Smooth endoplasmic reticulum
    Left column in sequence:
    Mitochondrion
    Secretory vesicle
    Nuclear membrane
    Nucleolus
    Centrioles
    Cytoplasm
    Cell membrane

11. It provides the surface through which substances enter and leave the cell and regulates passage through the membrane.

12. Oxygen diffuses from the high concentration in the alveoli of the lungs to the lower concentration in the blood. Carbon dioxide diffuses from the higher concentration in the blood into the lower concentration in the lungs.

13. Facilitated requires a carrier molecule and simple diffusion does not. Facilitated and simple diffusion are alike in that they both move substances down a concentration gradient and do not require energy.

14. The conditions include a selectively permeable membrane between two solutions of different concentrations. The solvent (water) moves from the lesser concentrated solution into the more concentrated solution.

15. a, Nothing will happen.
    b. The RBC will shrink or crenate.
    c. The RBC will swell and possibly rupture.

16. Pressure gradients direct the movement through a membrane.

17. Active transport is different from facilitated diffusion because it uses energy and moves substances against a concentration gradient, from lower concentration to higher concentration. They both involve carrier molecules and move solutes.

18. Endocytosis is taking material into the cell; exocytosis is removing material from the cell as in secretion. Phagocytosis takes solid particles into the cell; pinocytosis takes fluid droplets into the cell.

19. **Interphase** is the period between active cell divisions, a time of growth and metabolism. It is the longest period of the cell cycle. Just prior to cell division, the DNA, mitochondria, and centrioles replicates in interphase.
    Prophase is the beginning of nuclear division. The chromatin shortens and thickens to become chromosomes, centrioles move to opposite ends of the cell, spindle fibers form, and the nuclear membrane and nucleolus disappear.
    Metaphase is the phase when chromosomes align along the center of the cell.
    Anaphase is the time when the centromeres separate and spindle fibers shorten to pull chromatids toward the centrioles at opposite ends of the cell.
    Telophase is nearly the opposite of prophase. The chromosomes uncoil to become long filaments of chromatin; the nuclear membrane and nucleolus reappear. Cytokinesis occurs at this time to form two daughter cells.

20. a. 2 cells
    b. 46 chromosomes
    c. 4 cells
    d. 23 chromosomes

21. A gene is the portion of a DNA molecule that contains the information for making a particular protein. They are part of the DNA in chromatin.

22. Adenine pairs with thymine; cytosine pairs with guanine. During DNA replication, the hydrogen bonds between the bases break and new strands of DNA are formed. Because of the exact base pairing, the new strands that are formed are identical to the previous DNA.

23. mRNA is a long strand that is complementary to the coding strand of DNA. It takes the genetic information from the DNA in the nucleus into the cytoplasm. It contains the codons for amino acids. tRNA is a molecule with a specific sequence of three bases, called an anticodon, and a specific amino acid. It transfers the amino acid to a chain to form a protein.
24. Transcription is the process in the nucleus by which mRNA is formed with bases that are complementary to the coding DNA strand. It contains a "negative" of the genetic code. Translation is the process in the cytoplasm by which the "negative" code on mRNA is translated into a specific amino acid sequence by the action of tRNA.

☞ **Answers to Learning Exercises**

Structure of the Generalized Cell (Objectives 1-3)
1. Structural support, channels for passage of materials, receptor sites, carrier molecules, identification markers for immunity.
2. G (cell membrane)
   F (centriole)
   E (chromatin)
   L (cilia)
   I (golgi apparatus)
   H (mitochondria)
   J (nuclear pore)
   D (nucleolus)
   C (nucleus)
   A (ribosome)
   K (RER)
   B (SER)
3. Mitochondria
   Lysosome
   Golgi apparatus
   Ribosomes
   Centrioles
   Cilia
   Endoplasmic reticulum
   Chromatin
   Flagella
   Nucleolus
   Cytoplasm
   Cell membrane
   Nucleus
   Microfilaments
   Microtubules

Cell Functions (Objectives 11-24)
1. Osmosis
   Simple diffusion
   Facilitated diffusion

2. Active transport
   Endocytosis
   Passive transport
   Exocytosis
   Active transport
   Passive transport
   Endocytosis
3. A is hypertonic.
   A will increase in volume.
4. Anaphase, Prophase, Metaphase
5. Mitosis: Somatic, 2, 46
   Meiosis: Gametes, 4, 23
6. mRNA: AUG/UGU/UAC/AUU/CAA/AAC
   tRNA: UAC/ACA/AUG/UAA/GUU/UUG

☞ **Answers to Chapter Self-Quiz**

1. J
   I
   H
   C
   A
   E
   D
   F
   G
   B
2. a
3. b
4. a
5. Pinocytosis
6. B
   D
   C
   D
   A
   D
   E
7. d
8. a
9. d
10. d

☞ **Answers to Terminology Exercises**

Phagocytosis
Isotonic
Hydrophilic
Cytoplasm
Endocytosis

Cell drinking
Pertaining to the body
Study of cells
Dislike or hate water
Within the matter of the cell

E (presence of a tiny network)
C (excessive solute strength)
A (pulling apart phase)
D (little organs)
B (condition of taking something into cell)

☞ Answers to Fun and Games

1. Flagella
2. Anaphase
3. Thymine
4. Microfilament
5. Hydrophilic
6. Ribosomes
7. Isotonic
8. Cytokinesis
9. Osmosis
10. Across: Chromatin
10. Down: Centriole
11. Metaphase
12. Telophase
13. Across: Cilia
13. Down: Codon
14. Anticodon
15. Diffusion
16. Nucleolus
17. Across: Cytoplasm
17. Down: Crenate
18. Lysosomes
19. Mitosis
20. Golgi
21. Cytosine

☞ Quiz/Test Questions

Note: There are fifty multiple-choice questions for this chapter in the computerized test bank.

Name the following:

1. Two main structural components of the cell membrane.
   Answer: phospholipd and protein.

2. Granular organelle that functions in protein synthesis.
   Answer: ribosome.

3. Membranous organelle that functions in energy production.
   Answer: mitochondria.

4. Primary nucleic acid found in the nucleolus.
   Answer: ribonucleic acid (RNA).

5. Filaments in the nucleus that are composed of DNA.
   Answer: chromatin.

6. Foundation or cytoskeletal components of cilia and flagella.
   Answer: microtubules.

7. Organelles that move substances across the surface of the cell membrane.
   Answer: cilia.

8. Type of passive transport in which a carrier molecule is necessary.
   Answer: facilitated diffusion.

9. Process by which cells engulf solid particles.
   Answer: phagocytosis.

10. Transport process by which a cell loses water to a hypertonic solution.
    Answer: osmosis.

11. Longest phase of a cell cycle.
    Answer: interphase.

12. Process of nuclear division in somatic, or body, cells.
    Answer: osmosis.

13. Phase of nuclear division in which the spindle fibers shorten and pull the chromosomes toward the centrioles.
    Answer: anaphase.

14. Specific type of molecule that transcribes the genetic code and carries it to the cytoplasm.
    Answer: messenger RNA (mRNA).

15. Specific type of molecule that consists of an anticodon and an amino acid.
    Answer: transfer RNA (tRNA).

True/False Questions:

1. The outer and inner phosphate layers of the cell membrane are hydrophilic.
   Answer: True.

2. Lipids in the cell membrane act as receptor sites for hormone action.
   Answer: False; proteins act as receptor sites.

3. Individuals engaged in prolonged substance abuse are likely to have increased smooth endoplasmic reticulum, particularly in the liver.
   Answer: True; SER functions in the detoxification of drugs.

4. A cell that is placed in a hypotonic solution is likely to shrink or crenate.
   Answer: False; it is likely to swell up and rupture due to osmosis.

5. DNA and RNA are both found inside the nucleus.
   Answer: True; DNA is in the chromatin, RNA in the nucleolus.

Questions for this chapter continue on page 136.

4 Tissues and Membranes

☞ ## Key Terms/Concepts

Adipose Fat tissue.

Areolar Type of connective tissue that contains fibers and a variety of cells soft, loose, matrix; loose connective tissue.

Cardiac muscle Muscle tissue that is found only in the heart; involuntary striated muscle.

Cartilage A type of connective tissue in which cells and fibers are embedded in a semisolid gel matrix.

Chondrocyte Cartilage cell.

Collagenous fibers Connective tissue fibers that contain collagen to give strength to the tissue.

Connective tissue Most abundant and widespread tissue in the body; includes bone, cartilage, adipose, and various fibrous tissues.

Elastic fibers Connective tissue fibers that contain elastin and can be stretched; also called yellow fibers.

Epithelial tissue Tissue that covers the body and its parts; lines parts of the body; classified according to shape and arrangement.

Fibroblast Connective tissue that produces fibers.

Histology Branch of microscopic anatomy that studies tissues.

Lacuna Space or cavity; space that contains bone or cartilage cells; pleural, lacunae.

Macrophage Large phagocytic cell.

Mast cell Connective tissue cell that contain heparin and histamine.

Meninges Connective tissue membranes that cover the brain and spinal cord.

Mucous membrane Epithelial membrane that secretes mucus and lines body cavities that open directly to the exterior; also called mucosa.

Nervous tissue Specialized tissue found in nerves, brain, and spinal cord.

Neuroglia Supporting cells of nervous tissue; cells in nervous tissue that do not conduct impulses.

Neuron Nerve cell; conducting cell of nervous tissue.

Osseous tissue Bone tissue; rigid connective tissue.

Osteocyte Bone cell.

Osteon Structural unit of bone; Haversian system.

Serous membrane Epithelial membrane that secretes a serous fluid and lines the closed body cavities and covers the organs within those cavities; also called serosa.

Skeletal muscle Muscle that is under voluntary or willed control; also called voluntary striated muscle.

Smooth muscle Muscle tissue that is neither striated nor controlled voluntarily; also called visceral muscle.

Synovial membrane Connective tissue membrane that lines the cavities of the freely movable joints and secretes synovial fluid.

Tissue Group of similar cells specialized to perform a certain function.

☞ ## Chapter Objectives

Upon completion of this chapter the student should be able

1. Explain what is meant by a tissue.
2. Define the term histology.
3. List the four main types of tissues found in the body.
4. Describe epithelial tissues in terms of structure, location, blood supply, and mitotic capabilities.
5. Describe how epithelial tissues are classified according to cell shape and according to the number of layers.
6. Describe each of the following specific epithelial tissues and state at least one location for each: simple squamous, simple cuboidal, simple columnar, pseudostratified columnar, stratified squamous, and transitional.
7. Distinguish between exocrine and endocrine glands.
8. Give an example of a unicellular exocrine gland.
9. Distinguish between simple and compound glands; and between tubular and alveolar glands.

10. Distinguish between merocrine, apocrine, and holocrine glands. Give an example of each type.
11. Describe the general characteristics of connective tissues.
12. Distinguish between collagenous fibers and elastic fibers.
13. Name three types of connective tissue cells.
14. Describe the features and location of loose connective tissue, adipose, and dense fibrous connective tissue.
15. Describe the general characteristics of cartilage.
16. Distinguish between hyaline cartilage, fibrocartilage, and elastic cartilage. State at least one location for each.
17. Describe the general characteristics and structural features of osseous tissue.
18. Name the intercellular matrix and three types of cells found in blood.
19. Name two types of contractile proteins found in muscle tissue.
20. Distinguish between skeletal muscle, smooth muscle, and cardiac muscle in terms of structure, location, and control.
21. Name two categories of cells in nerve tissue.
22. Identify the three principal parts of a neuron.
23. Distinguish between epithelial membranes and connective tissue membranes.
24. Describe mucous membranes in terms of structure, secretions, and location.
25. Describe serous membranes in terms of structure, secretions, and location.
26. Identify the two layers found in serous membranes.
27. State the location of the pleura, pericardium, and peritoneum.
28. Name two connective tissue membranes and state a location for each one.
29. Identify the three layers of the meninges.

☞ Chapter Outline/Summary

Body Tissues (Objectives 1 -22)
- A tissue is a group of similar cells collected together by an intercellular matrix. Histology is the study of tissues. There are four main types of tissues in the body: epithelial, connective, muscle, and nerve tissue.

Epithelial tissue (Objectives 4 - 10)
- Epithelial tissues consist of tightly packed cells with little intercellular matrix, have one free surface, are avascular, and reproduce readily. They cover the body, line body cavities, and cover organs within body cavities. The cells may be squamous, cuboidal, or columnar in shape, and may be arranged in single or multiple layers.
- Simple squamous epithelium consists of a single layer of flat cells. Because it is so thin, this tissue is well suited for diffusion and filtration.
- Simple cuboidal epithelium consists of a single layer of cells shaped like a cube. This tissue is found in the kidney tubules where it functions in absorption and in glandular tissue where it functions in secretion.
- Simple columnar epithelium consists of a single layer of cells that are taller than they are wide. This tissue lines the stomach and intestines where it functions in secretion and absorption. Microvilli and goblet cells are frequently found in simple columnar epithelium.
- Pseudostratified columnar epithelium appears to be stratified but is not. All cells are attached to the basement membrane, but not all reach the surface. Cilia and goblet cells are often associated with this tissue. This tissue lines portions of the respiratory trace and some of the tubes of the reproductive tract.
- Stratified squamous epithelium consists of several layers of cells and the ones at the surface are flat, squamous cells. Its primary function is protection because it is thicker than other epithelia. This tissue forms the outer layer of the skin, and the linings of the mouth, anus, and vagina.
- Transitional epithelium has several layers but can be stretched in response to tension. This tissue is found in the lining of the urinary bladder.
- Glandular epithelium
  - Exocrine glands secrete their product onto a free surface through a duct.
  - Endocrine glands are ductless glands; they secrete their products into the blood.
  - Goblet cells are unicellular exocrine glands; other exocrine glands are multicellular.
  - The ducts of simple glands have no branches; compound glands have branched ducts; tubular glands have a constant diameter; acinar glands have a saccular distal end.
  - Merocrine glands lose no cytoplasm with their secretion; apocrine glands lose a portion of the cell with the secretory product; and in holocrine glands, the entire cell is discharged with the secretion.

Connective tissue (Objectives 11 - 18)
- Connective tissue has an abundance of intercellular matrix with relatively few cells. Strong and flexible collagenous fibers and yellow elastic fibers are frequently found in the matrix. Fibroblasts, macrophages, and mast cells are three of the most common connective tissue cells.
- Loose connective tissue is characterized by a loose network of collagenous and elastic fibers and a variety of connective tissue cells. The predominant cell is the fibroblast. It fills spaces in the body and binds structures together.
- Adipose is commonly called fat. It forms a protective cushion around certain organs, provides insulation, and is an efficient energy storage material.
- Dense fibrous connective tissue is characterized by densely packed collagenous fibers in the matrix. This tissue has a poor vascular supply. Tendons and ligaments are formed from dense fibrous connective tissue.
- Cartilage has the protein chondrin in the matrix; cells called chondrocytes are located in spaces called lacunae. Cartilage is typically surrounded by a fibrous connective tissue called perichondrium. Blood vessels do not penetrate cartilage so cellular reproduction and healing occur slowly.
  - Hyaline is the most common type of cartilage. It is found at the ends of long bones, in the trachea, costal cartilages, and in the fetal skeleton.
  - Fibrocartilage has an abundance of collagenous fibers in the matrix and is found in the intervertebral discs.
  - Elastic cartilage has an abundance of elastic fibers in the matrix and is found in the framework of the external ear.
- Bone, or osseous tissue, is a rigid connective tissue. Mineral salts in the matrix give strength and hardness to bone. The structural unit of bone is the osteon, or Haversian system. The Haversian canal, lamellae, osteocytes, lacunae, and canaliculi are all parts of the osteon.
- Blood is a connective tissue that has a liquid matrix called plasma. The cells found in blood are the erythrocytes, which transport oxygen; leukocytes, which fight disease; and thrombocytes, which function in blood clotting.

Muscle tissue (Objectives 19 and 20)
- Actin and myosin are contractile proteins in muscle tissue.
- Skeletal muscle fibers are cylindrical, multinucleated, striated, and under voluntary control. Skeletal muscles are attached to the skeleton.
- Smooth muscle cells are spindle-shaped, have a single, centrally located nucleus, lack striations, and are involuntary. Smooth muscle is found in the walls of blood vessels and internal organs.
- Cardiac muscle has branching fibers, one nucleus per cell, striations, intercalated disks, and are involuntary. Cardiac muscle is in the wall of the heart.

Nerve tissue (Objectives 20 and 21)
- Neurons and neuroglia are the cells found in nerve tissue.
- Neurons are the conducting cells of nerve tissue. They have a cell body with efferent processes called axons and afferent processes called dendrites.
- Neuroglia are the supporting cells of nerve tissue. They do not conduct nerve impulses.

Body Membranes (Objectives 23 - 29)
- Mucous and serous membranes are epithelial membranes formed from epithelial tissue and the connective tissue to which it is attached. Synovial membranes and the meninges are connective tissue membranes, which contain only connective tissue.
- Mucous membranes are epithelial membranes. They line body cavities that open to the outside, such as the mouth, stomach, intestines, urinary bladder, and respiratory tract. Mucous membranes secrete mucus for protection and lubrication.
- Serous membranes are epithelial membranes. They line body cavities that do not open to the outside and cover the organs within these cavities. Serous membranes always consist of two layers, the visceral layer and the parietal layer. Serous fluid is secreted between the two layers. The pleura is the serous membrane around the lungs, pericardium is around the heart, and peritoneum is in the abdomen.
- Synovial membranes are connective tissue membranes that line joint cavities. They s secrete synovial fluid into the joint cavity for lubrication.
- Meninges are connective tissue membranes around the brain and spinal cord. Dura mater is the outer layer of the meninges,

arachnoid is the middle layer, and pia mater is the innermost layer.

☞ **Answers to Review Questions**

1. A tissue is a group of cells that have similar structure and they function together as a unit.
2. Histology
3. Epithelial, connective, muscular, nervous
4. Functions: protection, secretion, absorption, excretion, filtration, diffusion, sensory perception.
   Location: covering body surfaces, lining body cavities and hollow organs, also in glands.
5. a. Relatively small amount of matrix
   b. Readily mitotic
   c. Has no blood vessels
   d. Closely packed cells
   e. Always has a free surface
6. Simple epithelium has one layer of cells; stratified has multiple layers.
7. a. Single layer of flattened cells.
   b. Single layer of tall cells.
   c. Tall cells of irregular height; looks stratified but is not.
   d. Multiple layers of cells with flattened cells at the surface.
8. a. Kidney tubules, thyroid gland, pancreas, salivary glands, covering the ovary.
   b. Lining the stomach and intestines.
   c. Respiratory tract and tubes of the male reproductive system.
   d. Outer layer of skin, lining of mouth, rectum, vagina.
9. Exocrine glands have ducts that carry secretion to a surface. Endocrine glands do not have ducts, but release their product into the blood which transports the product to the target cells.
10. Unicellular exocrine glands
11. Tubular glands have a constant diameter, with secretory portion the same size as duct; alveolar glands have an enlarged, rounded, saccular secretory portion.
12. In merocrine glands the product is released through the cell membrane and no cytoplasm is lost. In apocrine glands a portion of the cell is pinched off and released with the product. In holocrine glands the entire cell is discharged with the secretory product.
13. Connective tissue has relatively few cells with a large amount of intercellular matrix. Epithelium has a large number of cells with a small amount of intercellular material.
14. Collagenous fibers are composed of the protein collagen, are strong and flexible, but not very elastic. Elastic fibers contain the protein elastin, are not particularly strong, but are elastic.
15. Fibroblast cells produce fibers.
    Macrophages are large phagocytic cells that function in defense.
    Mast cells contain heparin, which inhibits blood clotting, and histamine, which functions in inflammation and in allergic reactions.
16. Areolar connective tissue; it attaches skin to underlying muscles and fills spaces in the body.
17. Dense fibrous connective tissue has closely packed bundles of collagenous fibers in the intercellular material. It is found in tendons and ligaments.
18. Cartilage has an abundance of a solid, but flexible, matrix that contains the protein chondrin. The cartilage cells, called chondrocytes, are enclosed within spaces, called lacunae, that are scattered throughout the matrix. Cartilage is surrounded by a membrane called the perichondrium.
19. Hyaline cartilage has relatively few fibers in the matrix; fibrocartilage contains an abundance of collagenous fibers in the matrix; elastic cartilage contains yellow elastic fibers in the matrix. Hyaline cartilage is found in the fetal skeleton, ends of long bones, and in costal cartilages. Fibrocartilage is found in the intervertebral disks, the symphysis pubis, and as pads in the knee.
20. The structural unit of bone is the osteon, or Haversian system. The osteon consists of a centrally located Haversian canal surrounded by concentric layers of matrix, which are called lamellae. The bone cells, or osteocytes, are located within spaces called lacunae, which are between the lamellae. Small channels, or canaliculi, radiate between lacunae and connect with the blood supply in the Haversian canal.
21. Red blood cells: erythrocytes
    White blood cells: leucocytes
    Platelets: thrombocytes
    Intercellular matrix of blood: plasma
22. Actin and myosin are the microfilaments in muscle tissue.
23. a. Skeletal muscle
    b. Cardiac muscle
    c. Smooth muscle
    d. Skeletal and cardiac muscle
    e. Visceral muscle
    f. Skeletal muscle

24. Neurons are the conducting cells of the nervous system; neuroglia are the supporting cells.
25. The cell body is the main part of the neuron and contains the organelles of the cell. The dendrites are afferent processes that carry impulses toward the cell body. An axon is an efferent process that carries impulses away from the cell body.
26. Epithelial membranes consists of epithelial tissue anchored to connective tissue. Connective tissue membranes contain only connective tissue.
27. Mucous membranes line body cavities that open to the outside. This includes the digestive tract, respiratory tract, urinary tract, and reproductive tract.
28. Serous membranes line the body cavities that do not open to the outside and they cover the organs within those cavities.
29. a. Lines the cavities that contain the lungs.
    b. Covers the heart; outermost layer of the heart.
    c. Lines the abdominal cavity.
30. a. Synovial membrane
    b. Meninges
    c. Meninges

☞ **Answers to Learning Exercises**

Body Tissues (Objectives 1-22)
1. A tissue is a group of similar cells collected together by an intercellular matrix.
2. Histology
3. Epithelial, connective, muscular, nervous
4. Epithelial tissues consist of <u>closely packed</u> cells with <u>very little</u> intercellular matrix. They have <u>one free</u> surface, have no <u>blood (vascular)</u> supply, and <u>reproduce</u> quickly. These tissues <u>cover</u> the body, <u>line</u> body cavities, and <u>cover</u> the organs within the body cavities. The cells may be <u>flat (squamous), cuboidal,</u> or <u>columnar</u> in shape.
5. Stratified squamous
   Skin

   Simple cuboidal
   Kidneys

   Pseudostratified ciliated columnar
   Respiratory passages
6. Endocrine
   Exocrine
   Exocrine
   Exocrine
   Endocrine

7. B
   C
   F
   G
   E
8. Collagenous (white)
   Macrophage
   Areolar (loose connective tissue)
   Adipose
   Dense fibrous
   Perichondrium
   Fibrocartilage
   Bone
   Haversian system (osteon)
   Osteocytes
   Blood (vascular)
   Erythrocytes
9. Areolar (loose)
   White collagenous (A)
   Yellow elastic (B)
10. Hyaline cartilage
    Chondrin (A)
    Chondrocyte (B)
    Lacuna (C)
11. Bone
    Lamellae (A)
    Haversian (osteonic) canal (B)
    Canaliculi (C)
12. Skeletal
    Sarcolemma (A)
    Actin and myosin
13. Smooth (visceral)
    Blood vessels and wall of hollow body organs
14. Cardiac
    Intercalated disc(B)
    Heart
15. Neurons
    Dendrites
    Axons
    Neuroglia

Body Membranes (Objectives 23-29)
1. Connective tissue membranes
   Mucous membranes
   Mucous membrane
   Serous membrane
   Parietal
   Visceral
   Pleura
   Pericardium
   Peritoneum
   Synovial
   Meninges
   Dura mater
   Arachnoid
   Pia mater

☞ **Answers to Chapter Self-Quiz**

1. A (closely packed cells)
   B (macrophages and fibroblasts)
   B (tendons and ligaments)
   A (simple and stratified
   D (axons and dendrites
   B (chondrocytes and osteocytes)
   C (specialized for contraction)
   B (blood and bones)
   A (has a free surface)
   C (actin and myosin)
2. K (outer layer of skin)
   H (kidney tubules)
   E (costal cartilage)
   F (attaches skin to underlying muscle)
   G (lining of stomach)
   E (most of the fetal skeleton)
   I (alveoli of the lungs)
   B (tendons)
   L (lining the urinary bladder)
   D (intervertebral discs)
3. Endocrine
4. d (unicellular glands)
5. e (collagenous fibers)
6. c (lacunae)
7. Thrombocyte
8. d (cardiac muscle)
9. Axons
10. a (mucous membranes)

☞ **Answers to Terminology Exercises**

Chondrocyte
Adenoma
Thrombocyte
Macrophage
Neuroglia

Without blood vessels
Produces fibers
Red cell
Layered
Flattened or scale-like in shape

E (appears to have layers but does not)
C (tumor of epithelial tissue)
A (fat tissue)
B (osseous tissue)
D (tumor of neuroglia cells)

☞ **Answers to Fun and Games**

Adipose (tissue filled with fat)
Neuron (conducts impulses)
Pleura (a serous membrane)
Chondrocyte (cartilage cell)

Epithelium (tissue that forms coverings)
Meninges (membranes around the brain)
Macrophage (large phagocytic cell)
Histamine (substance in mast cells)
Erythrocyte (red blood cell)
Areolar (loose connective tissue)
Visceral (smooth muscle)
Synovial (membrane in movable joints)
Squamous (flat epithelial cells)
Fibrocartilage (tissue in intervertebral discs)
Neuroglia (nerve glue)

☞ **Quiz/Test Questions**

Note: There are fifty multiple-choice questions for this chapter in the computerized test bank.

Name the following:

1. Term that denotes the study of tissues.
   Answer: histology.

2. Tissue with tightly packed cells and little intercellular material.
   Answer: epithelial.

3. Specific tissue that is good for diffusion in the alveoli of the lungs and capillary walls.
   Answer: simple squamous epithelium.

4. A unicellular exocrine gland.
   Answer: goblet cell.

5. Gland type with a bulbous or rounded secretory portion.
   Answer: alveolar or acinar.

6. Type of gland in which a portion of the cell is pinched off and released with the secretory product.
   Answer: apocrine.

7. Connective tissue cell that produced collagen fibers.
   Answer: fibroblast.

8. Connective tissue that is filled with fat.
   Answer: adipose.

9. Space in which a chondrocyte is located.
   Answer: lacuna.

10. Tissue type found in tendons and ligaments where strength is essential.
    Answer: dense fibrous connective tissue.

11. Type of tissue in the fetal skeleton and at the ends of long bones.
    Answer: hyaline cartilage.

12. Small channels that radiate from the osteocytes.
    Answer: canaliculi.

Questions for this chapter continue on page 136.

5 Integumentary System

Arrector pili Muscle associated with hair follicles.

Ceruminous gland A gland in the ear canal that produces cerumen or ear wax.

Cutaneous membrane Primary organ of the integumentary system; the skin.

Dermis Inner layer of the skin, which contains the blood vessels, nerves, glands, and hair follicles; also called stratum corium.

Epidermis Outermost layer of the skin.

Eponychium A fold of stratum corneum at the proximal border of the visible portion of a nail; also called cuticle.

Inflammation Localized protective response to tissue injury; characterized by redness, swelling, heat, and pain.

Integumentary Pertaining to the skin and related structures.

Keratinization Process by which the cells of the epidermis become filled with keratin and move to the surface where they are sloughed off.

Melanin A dark brown or black pigment found in parts of the body, especially skin and hair.

Sebaceous gland An oil gland, which produces sebum or body oil.

Subcutaneous layer Below the skin; A sheet of areolar connective tissue and adipose beneath the dermis of the skin; also called hypodermis or superficial fascia.

Sudoriferous gland A gland in the skin that produces perspiration; also called sweat gland.

☞ Chapter Objectives

Upon completion of this chapter the student should be able to:

1. Name the components of the integumentary system.
2. Name the two layers of the skin and a third, supporting, layer.
3. Describe the structure of the epidermis of the skin.
4. Describe the structure of the dermis of the skin.
5. Give another name for the subcutaneous layer and describe its structure.
6. Discuss three factors that influence skin color.
7. Describe the structure of hair and its relationship to the skin.
8. Describe the structure of nails and their relationship to the skin.
9. Discuss the characteristics and functions of sebaceous glands.
10. Distinguish between two types of sudoriferous glands on the basis of distribution and secretory product.
11. Specify the secretory product and location of ceruminous glands.
12. Discuss four functions of the integumentary system.

☞ Chapter Outline/Summary

Structure of the Skin (Objectives 1-5)
- The integumentary system includes the skin with its glands, hair, and nails. The skin has an outer epidermis and an inner dermis. These are anchored to underlying tissues by the hypodermis or subcutaneous tissue.

Epidermis (Objective 3)
- The epidermis is stratified squamous epithelium. In thick skin there are five distinct regions in the epidermis.
- The bottom layer is the stratum basale. It is closest to the blood supply, is actively mitotic, and contains melanocytes.
- The other layers are the stratum spinosum, stratum granulosum, stratum lucidum, and stratum corneum.
- The stratum lucidum is present only in thick skin.
- The outermost layer, the stratum corneum, is continually sloughed off and replaced by cells from deeper layers.

Dermis (Objective 4)
- The dermis is also called the stratum corium. It is composed of connective tissue with blood vessels, nerves, and accessory structures embedded in it.
- The dermis consists of two layers, the upper papillary layer and the deeper reticular layer.

Subcutaneous layer (Objective 5)
- The hypodermis anchors the skin to the underlying organs.
- The hypodermis has adipose in it, which acts as a cushion, as a heat insulator, and can be used as an energy source.

Skin Color (Objective 6)
- Basic skin color is due to the amount of melanin produced by the melanocytes in the stratum basale.
- Carotene, a yellow pigment, gives a yellow tint to the skin.
- Blood in the dermal blood vessels gives a pink color to the skin.

Epidermal Derivatives (Objectives 7-11)
- Hair, nails, sweat glands, and sebaceous glands are derived from the epidermis and are embedded in the dermis.

Hair and hair follicles (Objective 7)
- The central core of hair is the medulla, which is surrounded by the cortex and the cuticle.
- Hair is divided into the visible shaft and the root, which is embedded in the skin and surrounded by the follicle.
- The distal end of the hair follicle expands to form a bulb around a central papilla.
- Stratum basale cells in the bulb undergo mitosis to form hair. This increases the length of the hair.
- Hair color is determined by melanocytes in the stratum basale.
- The cross sectional shape of the hair shaft determines if hair is straight or curly.
- Arrector pili muscles contract in response to cold and fear. This makes the hair "stand on end."

Nails (Objective 8)
- Nails are thin plates of keratinized stratum corneum.
- Each nail has a free edge, nail body, and a nail root.
- Other structures associated with the nail are the eponychium, nail bed, nail matrix, and lunula.
- Nails are derived from the stratum basale in the nail bed.

Glands (Objectives 9-11)
- Sebaceous glands are oil glands and are associated with hair follicles. These glands secrete sebum, which helps keep hair and skin soft and pliable and helps prevent water loss.
- Sweat glands are called sudoriferous glands. Merocrine sweat glands open to the surface of the skin through sweat pores and secrete

perspiration in response to nerve stimulation and in response to heat. Apocrine sweat glands are larger than the merocrine glands and their distribution is limited to the axillae and external genitalia. Ducts of apocrine sweat glands open into hair follicles. Apocrine sweat glands become active at puberty and are stimulated in response to pain, emotional stress, and sexual arousal.
- Ceruminous glands are modified sweat glands found only in the external auditory canal. These glands secrete cerumen, or earwax.

Functions of the Skin (Objective 12)
- The skin protects against water loss, invading organisms, ultraviolet light, and other injuries.
- Sense receptors in the skin detect information about the environment and also serve as a means of communication between individuals.
- The skin functions in the regulation of body temperature. Constriction and dilation of blood vessels affects the amount of heat that escapes from the skin into the surrounding air. Sweat glands are stimulated in response to heat and inactive in cold temperatures. Adipose in the subcutaneous tissue helps insulate the body.
- Precursors for vitamin D are found in the skin. When the skin is exposed to ultraviolet light, these precursors are converted into active vitamin D.

☞ **Answers to Review Questions**

1. Skin, glands, hair, nails
2. Epidermis: outer layer of skin, composed of stratified squamous epithelium.
   Dermis: inner layer of skin, composed of connective tissue with epidermal derivatives embedded in it.
   Hypodermis: anchors the skin to underlying tissues, also called subcutaneous tissue or superficial fascia.
3. Stratum corneum: surface region of the epidermis composed of 20 to 30 layers of flattened, dead, keratinized cells.
   Stratum lucidum: a clear or translucent band of cells just beneath the stratum corneum; found only in thick skin.
   Stratum granulosum: Two or three layers of flattened cells where keratinization begins so the cells appear granular.
   Stratum spinosum: Several layers of cells beneath the stratum granulosum; cells have spiny projections and have limited mitotic ability.

Stratum basale: Bottom row of cells, columnar in shape and next to the dermis; actively mitotic layer; also contains melanocytes.

4. Upper papillary layer has numerous projections, or papillae, that extend into the epidermis. These contain blood vessels, nerve endings, and sensory receptors.
Lower reticular layer is thicker than the papillary layer and has an abundance of connective tissue fibers to provide strength and resistance.

5. Hypodermis anchors the skin to underlying organs. Adipose in the hypodermis acts as a cushion against mechanical shock, as an insulator in temperature regulation, and can be used as an energy source.

6. Amount of melanin is the primary factor that influences skin color. Two other factors are the yellow pigment carotene and the blood in the dermal blood vessels.

7. Stratum basale of the epidermis.

8. The **shaft** of the hair is the visible portion and the **root** is the portion embedded in the skin. The central core of the hair is the **medulla**, which is surrounded by several layers of cells that are the **cortex**. This is covered by a single layer of overlapping keratinized cells called the **cuticle**. The root is enclosed in a **hair follicle**, which expands at the base to form a **hair bulb**. This surrounds a **papilla**, which is a projection of dermis with blood vessels. An **arrector pili** muscle is attached to the root.

9. The visible portion of a nail is the **nail body**. The distal end is the **free edge** and the portion that is covered with skin is the **nail root**. The fold of skin that covers the proximal portion of the nail body is the **eponychium**. Stratum basale from the epidermis grows under the nail body to form the **nail bed**. This is thickened at the proximal end to form the **nail matrix**. The portion of the nail body that is over the matrix appears lighter in color and is called the **lunula**.

10. Sebaceous glands are generally associated with hair follicles and open into the follicles. They secrete sebum to keep hair and skin soft and pliable. Sebum also helps prevent fluid loss.

11. Sudoriferous glands are sweat glands. **Merocrine** glands are more numerous and more widely distributed than apocrine glands. Their secretion is a watery fluid with a few salts, which is secreted in response to increased temperature and emotional stress.

Apocrine glands are larger than merocrine and are limited to the axillae and external genitalia where the ducts open into hair follicles. Secretory product contains organic compounds and is released in response to pain, emotional stress, and sexual arousal.

12. **Ceruminous glands** are modified sweat glands in the external auditory canal. They secrete a waxy substance called cerumen.

13. Protection against water loss, ultraviolet light, invading organisms, and injuries; sensory reception; temperature regulation; vitamin D synthesis.

14. The skin functions in temperature regulation through the constriction and dilation of cutaneous blood vessels and the activity of the sweat glands. When temperature increases, blood vessels dilate to bring more blood to the surface so the heat can be dissipated. When temperature decreases, the vessels constrict to retain heat in the body. When temperature increases the sweat glands become more active and the evaporation of the perspiration cools the body.

☞ **Answers to Learning Exercises**

Structure of the Skin (Objectives 1-5)
1. Skin, hair, nails, and glands.
2. Hypodermis
   Stratum basale
   Stratum basale
   Stratum germinativum
   Dermis
   Stratum spinosum
   Hypodermis
   Stratum lucidum
   Stratum granulosum
   Stratum corneum
   Dermis
   Dermis
   Hypodermis
   Stratum basale
   Dermis
3. C (arrector pili muscle)
   H (blood vessel)
   A (epidermis)
   G (hair bulb)
   F (sebaceous gland)
   B (stratum basale)
   E (stratum corneum)
   D (sweat gland)

Skin Color (Objective 6)
1. Melanin
2. Carotene
3. Blood vessels in the dermis

4. As the cells with increased melanin are pushed to the surface, die, and are sloughed off, the "tan" lightens because the new cells do not have as much melanin.

Epidermal Derivatives (Objectives 7-11)

1. Follicle
   Arrector pili
   Stratum basale
   Medulla
   Lunula
   Sebaceous
   Ceruminous
   Merocrine
   Sudoriferous
   Bulb
   Eponychium
   Apocrine

Functions of the Skin (Objective 12)

1. Protection, sensory reception, regulation of body temperature, synthesis of vitamin D
2. Keratin
3. Oily secretions of sebaceous glands
4. Melanin
5. Dermis
6. Blood vessels dilate to bring more blood to surface to radiate heat from the body. In cold, they constrict to conserve heat inside the body.
   In heat, sweat glands are actively producing perspiration which carries large quantities of heat to the surface. Evaporation then cools the body. In cold, sweat glands are inactive.
7. Precursors for vitamin D are in the skin and when exposed to ultraviolet from the sun, vitamin D is formed.

☞ **Answers to Chapter Self-Quiz**

1. B (outermost layer)
   A (responsible for fingerprints)
   B (has 5 distinct layers)
   C (contains adipose)
   B (has cells responsible for skin color)
   B (stratified squamous epithelium)
   B (hair and nails derived from this layer)
   A (has sense receptors and hair embedded)
2. a
3. e
4. b
5. True

☞ **Answers to Terminology Exercises**

Melanoma
Hyponychium
Onychiectomy

Hypodermis
Melanocyte

Thick skin
Below the skin
Surgical repair of the skin
Study of the skin
Lack of sweat

C (clear layer of the skin)
E (plastic surgery for removal of wrinkles)
B (condition of dry, scaly skin)
A (ear wax)
D (condition of thick nails)

☞ **Answers to Fun and Games**

Across

1. Reticular
7. Corneum
12. Bee
13. Rim
14. MO
15. Eons
17. Carotene
19. Seek
21. PA
22. Ride
23. Up
24. Dermis
28. Edging
31. Sr
32. Aces
34. AP
36. Su
37. Irons
40. Awl
42. Big
43. Sal
45. Melanocyte
46. Us
47. Apple
49. En
50. Gem
53. Linens
54. PNS
56. LP
57. El
58. At
59. Blimp
62. ICUs
64. Shoo
65. Ac
67. Den
68. Day
70. Subcutaneous

75. OR
76. Dam
78. Tutor
79. LP
80. Hypodermis
83. At
84. Meal
86. Il
87. Lain
89. Axis
90. Striae
92. Illest
94. Skin
95. BOA
96. ND
97. UE
98. Opt
99. Lucidum
101. Cue
102. Femme
104. SOS
105. Any
106. Fascia
107. Old

DOWN

1. Rear
2. Eerie
3. Integument
4. Urn
5. Lie
6. AM
7. Coeds
8. Re
9. Normal
10. En
11. US
12. BC
14. ME
16. Dad
18. Odds
19. Spinosum
20. Keratin
25. IC
26. Sebum
27. Basale
29. Nile
30. Granulosum
33. Sis
35. Papillary
38. ON
39. Sc
41. We
44. LPN
48. Leak
51. Epidermis
52. CPS

55. Sweat
56. Lunula
59. BH
60. IOU
61. Pac
63. CEO
66. Cutaneous
68. Doha
69. Radii
71. Basale
72. Tut
73. No
74. Splinted
76. Do
77. Melanin
81. Petals
82. Ilium
85. Eskimo
88. IL
89. ATT
90. SO
91. Eddy
93. Spec
95. Bus
100. Ca
101. Ca
102. FA
103. Ml

☞ **Quiz/Test Questions**

Note: There are fifty multiple-choice questions for this chapter in the computerized test bank.

Name the following:

1. Uppermost layer of the epidermis.
   Answer: stratum corneum.

2. Layer of epidermis that is present only in thick skin.
   Answer: stratum lucidum.

3. Layer of the epidermis that contains the melanocytes.
   Answer: stratum basale.

4. Layer that attaches skin to the underlying tissues such as muscle.
   Answer: hypodermis or subcutaneous layer.

5. Uppermost layer of the dermis.
   Answer: papillary layer.

6. Pigment that creates brown tones in the skin.
   Answer: melanin.

7. Muscle associated with hair.
   Answer: arrector pili.

Questions for this chapter continue on page 136.

6 Skeletal System

☞ Key Terms/Concepts

Amphiarthrosis A slightly movable joint; plural, amphiarthroses.

Appendicular skeleton Bones of the upper and lower extremities of the body.

Appositional growth Growth due to material deposited on the surface, such as the growth in diameter of long bones.

Articulation A joint; a point of contact between bones.

Axial skeleton Bones of the head, neck, and torso.

Cancellous bone Bone that consists of numerous branching bony plates with interconnecting spaces between the plates; spongy bone.

Compact bone Bone that consists of tightly packed osteons with little or no space between them.

Diaphysis The long straight shaft of a long bone.

Diarthrosis Freely movable joint characterized by a joint cavity; also called a synovial joint; plural diarthroses.

Endochondral ossification Method of bone formation in which cartilage is replaced by bone.

Epiphyseal plate The cartilaginous plate between the epiphysis and diaphysis of a bone, which is responsible for the lengthwise growth of a long bone.

Epiphysis The end of a long bone.

Hematopoiesis Blood cell production, which occurs in the red bone marrow; also called hemopoiesis.

Medullary cavity Space in the shaft of a long bone; contains yellow marrow.

Intramembranous ossification Method of bone formation in which the bone is formed directly in a membrane.

Ossification Formation of bone; also called osteogenesis.

Osteoblast Bone forming cell.

Osteoclast Cell that destroys or resorbs bone tissue.

Osteocyte Mature bone cell.

Osteon Structural unit of bone; Haversian system.

Sesamoid bone A small bone, usually found in a tendon.

Synarthrosis An immovable joint; plural, synarthroses.

Trabeculae Thin plates of bone tissue, arranged in an irregular latticework, found in spongy bone.

Wormian bone A small bone located within a suture between certain cranial bones; also called sutural bone.

☞ Chapter Objectives

Upon completion of this chapter the student should be able to:

1. Discuss five functions of the skeletal system.
2. Distinguish between compact and spongy bone on the basis of structural features.
3. Classify bones according to size and shape.
4. Identify the general features of a long bone.
5. Define osteogenesis.
6. Identify three types of cells involved in bone formation and remodeling.
7. Distinguish between intramembranous and endochondral ossification.
8. Describe the processes by which bones increase in length and in diameter.
9. Distinguish between the axial and appendicular skeletons, and state the number of bones in each.
10. Identify the bones of the skull and their important surface markings.
11. Describe and identify the hyoid bone.
12. Describe the curvatures of the vertebral column.
13. Identify the general structural features of vertebrae.
14. Compare cervical, thoracic, lumbar, sacral, and coccygeal vertebrae and state the number of each type.
15. Identify the structural features of the ribs and sternum.
16. Distinguish between true ribs and false ribs.
17. Identify the features of each bone of the pectoral girdle.
18. Identify the bones of the upper extremity and the major features of each bone.
19. Identify the features of the pelvic girdle.

20. Distinguish between the false pelvis and the true pelvis.
21. Identify the bones of the lower extremity and the major features of each bone.
22. Compare the structure and function of three types of joints.

☞ **Chapter Outline/Summary**

Overview of Skeletal System (Objectives 1-9)
Functions of the skeletal system (Objective 1)
- Bones support the soft organs of the body and support the body against the pull of gravity.
- Bones protect soft body parts such as the brain and heart.
- Bones work with muscles to produce movement.
- Bones store minerals, especially calcium.
- Most blood cell formation, hematopoiesis, occurs in red bone marrow.

Structure of bone tissue (Objective 2)
- The microscopic unit of compact bone is the osteon (Haversian system), which consists of an osteonic canal, lamellae of matrix, osteocytes in lacunae, and canaliculi.
- In compact bone the osteons are packed closely together.
- Spongy bone is less dense than compact bone and consists of bone trabeculae around irregular cavities that contain red bone marrow. Trabeculae are organized to provide maximum strength to a bone.

Classification of bones (Objective 3)
- Long bones are longer than they are wide. An example is the femur in the thigh.
- Short bones are roughly cube shaped. Examples include the bones in the wrist.
- Flat bones have inner and outer tables of compact bone with a diploe of spongy bone in the middle. Examples include the bones of the cranium.
- Irregular bones are primarily spongy bone with a thin layer of compact bone. Vertebrae are irregular bones.

General features of a long bone (Objective 4)
- Long bones have a diaphysis, around a medullary cavity, with an epiphysis at each end.
- The epiphysis of a long bone is covered by articular cartilage.
- Except in the region of the articular cartilage, long bones are covered by periosteum and lined with endosteum.
- All bones have surface markings that make each one unique.

Bone development and growth (Objectives 5 - 8)
- Bone development is called osteogenesis.
- Osteoblasts, osteocytes, and osteoclasts are three types of cells involved in bone formation and remodeling.
- Intramembranous ossification involves the replacement of connective tissue membranes by bone tissue. Flat bones of the skull develop this way.
- Most bones develop by endochondral ossification. In this process, the "bones" first form as hyaline cartilage models, which are later replaced by bone.
- Long bones increase in length at the cartilaginous epiphyseal plate. When the epiphyseal plate completely ossifies, increase in length is no longer possible.
- Increase in the diameter of long bones occurs by appositional growth. Osteoclasts break down old bone next to the medullary cavity at the same time osteoblasts form new bone on the surface.

Divisions of the skeleton (Objective 9)
- The adult human skeleton has 206 named bones in addition to varying numbers of wormian and sesamoid bones.
- The axial skeleton, with 80 bones, forms the vertical axis of the body. It includes the head, vertebrae, ribs, and sternum.
- The appendicular skeleton, with 126 bones, includes the appendages and their attachments to the axial skeleton.

Axial Skeleton (Objectives 10-16)
- There are 80 bones in the axial skeleton, which consists of the skull, hyoid, vertebral column, and thoracic cage.
Skull (Objective 10)
- The skull includes the bones of the cranium, face, and the auditory ossicles.
- With the exception of the mandible, the cranial and facial bones are joined by immovable joints called sutures.
- The cranial bones are frontal (1), parietal (2), temporal (2), occipital (1), ethmoid (1), and sphenoid (1).
- The frontal, ethmoid, and sphenoid bones contain cavities called paranasal sinuses.
- The bones of the face are maxillae (2), nasal (2), lacrimal (2), zygomatic (2), vomer (1), inferior nasal conchae (2), palatine (2), and mandible (1).
- Each maxilla contains a large maxillary sinus.
- Six auditory ossicles, three in each ear, are located in the temporal bones.

Hyoid (Objective 11)

- The hyoid bone is a U-shaped bone in the neck. It does not articulate with any other bone and functions as a base for the tongue and as an attachment for muscles.

Vertebral column (Objectives 12 - 14)

- The vertebral column contains 26 vertebrae that are separated by intervertebral discs.
- Four curvatures add strength and resiliency to the column. The thoracic and sacral curvatures are concave anteriorly; the cervical and lumbar curvatures are convex anteriorly.
- All vertebrae have a body or centrum, vertebral arch, vertebral foramen, transverse processes, and a spinous process.
- Vertebrae are classified as cervical (7), thoracic (12), lumbar (5), sacral (5 fuse to 1), and coccygeal (4 fuse to 1).
- Cervical vertebrae have bifid spinous processes and transverse foramina. The first two cervical vertebrae are unique. The atlas, C_1 is a ring that holds up the occipital bone. The axis, C_2, has a dens or odontoid process.
- Thoracic vertebrae have facts on the bodies and transverse processes for articulation with the ribs.
- Lumbar vertebrae have large heavy bodies to support body weight.
- The sacrum is formed from 5 separate bones that fuse together in the adult.
- The coccyx is the most distal part of the vertebral column. Three to five bones fuse to form the coccyx.

Thoracic cage (Objectives 15 and 16)

- The thoracic cage consists of the thoracic vertebrae, the ribs, and the sternum.
- The sternum has three parts: the manubrium, the body, and the xiphoid. The jugular notch in the manubrium is an important landmark. The sternal angle is where the manubrium joins the body.
- There are 7 pairs of true ribs (vertebrosternal ribs), and 5 pairs of false ribs. The upper 3 pairs of false ribs are vertebrochondral, the lower 2 pairs are vertebral or floating ribs.

Appendicular Skeleton (Objectives 17-20)

- The appendicular skeleton consists of the appendages and their attachments to the axial skeleton. There are 126 bones in this division of the skeleton.

Pectoral girdle (Objective 17)

- The pectoral girdle is formed by 2 clavicles and 2 scapulas.
- The pectoral girdle supports the upper extremities.

Upper extremity (Objective 18)

- The upper extremity includes the humerus, radius, ulna, carpus, metacarpus, and phalanges.
- The humerus is the bone in the arm, or brachium.
- The radius is the lateral bone in the forearm and the ulna is the medial bone.
- The hand is composed of a wrist, palm, and five fingers. The wrist, or carpus, contains 8 small carpal bones; the palm, or metacarpus, contains 5 metacarpal bones; the 14 bones in the fingers are phalanges.

Pelvic girdle (Objectives 19 and 20)

- The pelvic girdle attaches the lower extremities to the axial skeleton and provides support for the weight of the body.
- The pelvic girdle consists of two ossa coxae, or innominate bones.
- Each os coxa is formed by 3 bones fused together: ilium, ischium, and pubis.
- The two ossa coxae meet anteriorly at the symphysis pubis.
- The ilium, ischium, and pubis meet in a large cavity, the acetabulum, that provides articulation for the femur.
- The false pelvis (greater pelvis) is the area between the flared portions of the ilium bones; the true pelvis (lesser pelvis) is inferior to the false pelvis and begins at the pelvic brim.

Lower extremity (Objective 21)

- The bones in the lower extremity are the femur, tibia, fibula, patella, tarsal bones, metatarsals, and phalanges.
- The bone in the thigh is the femur. The head of the femur articulates in the acetabulum of the os coxa. Distally, the femur articulates with the tibia.
- The medial bone in the leg is the tibia. Proximally it articulates with the femur, distally with the talus. The lateral bone in the leg is the fibula.
- Seven tarsal bones form the ankle. The talus is the tarsal bone that articulates with the tibia. The calcaneus is the heel bone.
- The bones in the instep of the foot are the five metatarsals.
- There are 14 phalanges in the toes of each foot.
- The patella is a sesamoid bone in the anterior portion of the knee joint.

Articulations (Objective 22)

- Synarthroses are immovable joints. An example is a suture in the skull.

- Amphiarthroses are slightly movable joints. The bones are connected by hyaline cartilage or fibrocartilage. Examples include the symphysis pubis and intervertebral discs.
- Diarthroses are freely movable joints. In these joints, the bones are held together by a fibrous joint capsule that is lined with a synovial membrane. These joints are sometimes called synovial joints. There are several types of freely movable joints: gliding, condyloid, hinge, saddle, pivot, and ball-and-socket.

☞ Answers to Review Questions

1. Support, protection, movement, mineral storage, blood cell formation
2. Compact bone is more dense than spongy bone. Compact bone consists of tightly packed osteons (Haversian systems). It is found in the diaphysis of long bones and in the outer and inner layers of flat bones. Spongy bone consists of plates of bone, called trabeculae, around irregular spaces filled with bone marrow. Spongy bone is found at the ends of long bones, the middle layer of flat bones, in short bones, and in irregular bones.
3. Long bones: femur, humerus, metacarpals, metatarsals, radius, ulna, tibia, fibula, phalanges.
   Short bones: carpals (wrist) and tarsals (ankle).
   Flat bones: bones of the cranium.
   Irregular bones: vertebrae, ethmoid, sphenoid.
4. Should include epiphysis, diaphysis, articular cartilage, region of spongy bone, region of compact bone, medullary cavity, periosteum, and endosteum.
5. Osteogenesis or ossification.
6. Osteoblasts are bone-forming cells.
   Osteocytes are mature bone cells.
   Osteoclasts break down and reabsorb bone.
7. In intramembranous ossification bone tissue develops within a connective tissue membrane and replaces it. In endochondral ossification bone tissue replaces hyaline cartilage models.
8. Bones increase in length at the epiphyseal plate.
9. Bones increase in diameter at the diaphysis. Osteoblasts lay down bone around the periphery while osteoclasts reabsorb bone from the inside, adjacent to the medullary cavity.

This keeps the bones from becoming too bulky and heavy.

10. Axial skeleton: 80 bones
    Appendicular skeleton: 126 bones
11. Frontal, parietal (2), temporal (2), occipital, ethmoid, sphenoid.
12. Maxillae (2), nasal (2), lacrimal (2), inferior nasal conchae (2), zygomatic (2), palatine (2), vomer (1), mandible (1).
13. Auditory ossicles are three tiny bones located in the middle ear cavity of each temporal bone.
14. The hyoid bone does not articulate directly with any other bone, but is suspended under the mandible and anchored by ligaments to the styloid process of the temporal bone.
15. Thoracic and sacral curvatures are present at birth and are concave anteriorly. The cervical curve develops when the infant begins to hold its head erect. The lumbar curve develops when the infant begins to stand and walk. The cervical and lumbar curves are convex anteriorly. The curves add strength and resilience to the vertebral column.
16. Body or centrum, vertebral arch, vertebral foramen, transverse processes, spinous processes.
17. Cervical vertebrae: (7) transverse foramen in the transverse processes and bifid (forked) spinous processes.
    Thoracic vertebrae: (12) articular facets for ribs and long, pointed spinous processes.
    Lumbar vertebrae: (5) large, heavy bodies and short, blunt spinous processes.
18. Sacrum is a triangular bone just below the lumbar vertebrae; consists of 5 bones that fuse into a single bone in the adult. It has superior articular facets, a sacral canal, an auricular surface, sacral foramina, and spinous tubercles. The coccyx is the terminal portion of the vertebral column. It consists of 3-5 coccygeal vertebrae that fuse into a single bone in the adult.
19. Manubrium, body, xiphoid process.
20. Vertebral ribs articulate with the vertebral column but have no anterior articulation. They are floating ribs. Each vertebrosternal rib articulates posteriorly with the vertebral column and anteriorly with the sternum by way of its individual costal cartilage. Vertebrochondral ribs articulate posteriorly with the vertebral column but anteriorly their costal cartilages attach to the costal cartilage of the previous rib.
21. Scapula and clavicle

22. Humerus: arm or brachium
    Radius: lateral side of forearm
    Ulna: medial side of forearm
    Carpals: 8 bones in the wrist
    Metacarpals: 5 bones in the hand
    Phalanges: 14 bones of the fingers
23. Ilium, ischium, pubis
24. **Symphysis pubis:** where the two pubis bones meet anteriorly.
    Acetabulum: large depression where the ilium, ischium, and pubis meet; where the head of the femur articulates.
    Iliosacral joint: where the ilium and sacrum meet.
    Obturator foramen: large opening between the pubis and ischium that serves as a passageway for blood vessels, nerves, and muscle tendons.
    Iliac crest: superior margin of the wing or ala of the ilium.
    Greater sciatic notch: deep indentation in the posterior region of the ilium.
    Ischial tuberosity: large, rough , inferior portion of the ischium.
25. False pelvis is the upper, larger region surrounded by the flared portions of the ilium bones and the lumbar vertebrae. The true pelvis (lesser pelvis) is the region below the pelvic brim.
26. Femur: thigh
    Fibula: lateral side of the leg
    Tibia: medial side of the leg
    Patella: sesamoid bone in anterior knee
    Tarsals: 7 bones in the ankle
    Metatarsals: 5 bones in the foot
    Phalanges: 14 bones in the toes
27. Synarthrosis: immovable, sutures of skull.
    Amphiarthrosis: slightly movable, symphysis pubis and intervertebral joints.
    Diarthrosis: freely movable, all synovial joints.
28. Articular cartilage, joint capsule, joint cavity, synovial membrane, synovial fluid.
29. Menisci are fibrocartilaginous pads in the knee that act as shock absorbers. Bursae are sacs filled with synovial fluid that act as cushions to reduce friction.

☞ Answers to Learning Exercises

Overview of Skeletal System (Objectives 1-9)

1. Support, protection, movement, storage, blood cell formation.
2. C (closely packed osteons)
   S (contains red bone marrow)
   S (trabeculae)
   C (canaliculi radiate from lacunae)
   S (contains irregular spaces)
3. F (has diploe of spongy bone)
   L (vertical dimension longer than horizontal)
   S (roughly cube shaped)
   S (primarily spongy bone)
   S (bones in wrist and ankle)
   L (bones in thigh and arm)
   F (spongy bone between two compact layers)
4. Diaphysis
   Epiphysis
   Periosteum
   Medullary cavity
   Articular cartilage
   Endosteum
5. Condyle
   Foramen
   Fossa
   Sinus
   Facet
   Trochanter
6. Osteogenesis/ossification
   Osteoblasts
   Osteocytes
   Osteoclasts
   Intramembranous
   Endochondral
   Epiphyseal plate
   Osteoblasts
   Osteoclasts
   Endochondral
7. 206 (complete skeleton)
   80 (axial skeleton)
   126 (appendicular skeleton)

Axial Skeleton (Objectives 10-16)

1. Frontal (1), parietal (2), occipital (1), temporal (2), ethmoid (1), sphenoid (1).
2. Nasal (2), maxillae (2), zygomatic (2), inferior nasal conchae (2), lacrimal (2), palatine (2), mandible (1), vomer (1).
3. Occipital (foramen magnum)
   Temporal (auditory meatus)
   Frontal (supraorbital foramen)
   Temporal (mastoid process)
   Sphenoid (optic foramen)
   Ethmoid (cribriform plate)
   Sphenoid (sella turcica)
   Mandible (ramus)
4. Hyoid
5. Atlas (first cervical vertebra)
   Thoracic (articulate with ribs)
   Body/centrum (weight bearing portion)
   Manubrium (superior portion of breastbone)
   Axis (second cervical vertebra)
   Cervical (type of vertebrae in neck)

Lumbar (heavy bodies and blunt processes)
Sternum (another name for breastbone)
Intervertebral discs (cartilaginous pads)

6. 7 pr=14 (true ribs)
   7 (cervical vertebrae)
   5 pr=10 (false ribs)
   12 (thoracic vertebrae)
   5 (lumbar vertebrae)
   7 pr=14 (vertebrosternal ribs)
   2 pr=4 (vertebral ribs)
   3 pr=6 (vertebrochondral ribs)

7. R (mandible)
   L (frontal)
   D (temporal)
   N (nasal)
   Q (maxilla)
   E (occipital)
   P (zygomatic)
   M (sphenoid)
   O (lacrimal)
   A (parietal)
   H (styloid process)
   F (auditory meatus)
   G (mastoid process
   I (mandibular condyle)
   J (ramus)
   K (coronal suture)
   B (squamosal suture)
   C (lambdoidal suture)

Appendicular Skeleton (Objectives 17-21)

1. Clavicle (anterior bone of pectoral girdle)
   Scapula (posterior bone of pectoral girdle)
   Scapula (has acromion and spine)
   Humerus (large bone in arm)
   Radius (bone on lateral side of forearm)
   Ulna (bone on medial side of forearm)
   Carpals (wrist bones)
   Metacarpals (form palm of the hand)
   Phalanges (form fingers and toes)
   Ulna (articulates with trochlea of humerus)
   Radius (articulates with capitulum)
   Olecranon process (projection on ulna)

2. Ilium
   Ischium (these three any order)
   Pubis
   Acetabulum
   Condyles
   Tibia
   Fibula
   Calcaneus
   Talus
   Patella
   Sacrum
   Femur

3. Male pelvis:
      Arch less than 90º
      Inlet narrow, heart-shaped
      Cavity narrow, deep, funnel-shaped
   Female pelvis:
      Arch more than 90º
      Inlet wider and oval
      Cavity broad, oval, shallow

4. The portion of the pelvis between the flared wings of the ilium bones is called the **false** pelvis. The portion inferior to the pelvic brim is the **true** pelvis.

5. A Frontal
   B Humerus
   C Os coxa (ilium)
   D Carpals
   E Femur
   F Patella
   G Metatarsals
   H Mandible
   I Clavicle
   J Sternum
   K Radius
   L Phalanges
   M Tibia
   N Phalanges

6. A Parietal
   B Scapula
   C Humerus
   D Ulna
   E Sacrum
   F Metacarpals
   G Fibula
   H Talus
   I Tarsals
   J Occipital
   K Cervical vertebrae
   L Ribs
   M Lumbar vertebrae
   N Radius
   O Femur
   P Tibia
   Q Calcaneus

Articulations (Objective 22)

1. Synarthrosis (sutures)
   Amphiarthrosis (slightly movable)
   Diarthrosis (meniscus)
   Synarthrosis (immovable)
   Diarthrosis (elbow and knee)
   Diarthrosis (ball and socket)
   Amphiarthrosis (ribs to sternum)
   Diarthrosis (may have bursae)
   Diarthrosis (joint capsule)
   Amphiarthrosis (symphysis pubis)

☞ **Answers to Chapter Self-Quiz**

1. B (carpals)
   A (femur)
   A (metatarsals)
   C (occipital)
   A (phalanges)
   D (sphenoid)
   B (tarsals)
   C (temporal)
   A (tibia)
   D (vertebrae)
2. A (shaft)
   E (hollow space in center)
   C (expanded ends)
   D (growth region)
   E (location of yellow marrow)
   C (location of red marrow)
   F (outer covering)
   B (lining on inside of shaft)
3. P (calcaneus)
   P (carpals)
   P (clavicle)
   X (ethmoid)
   P (femur)
   P (humerus)
   X (lumbar vertebrae)
   X (nasal conchae)
   X (mandible)
   X (occipital)
   P (os coxa)
   P (patella)
   X (ribs)
   X (sacrum)
   P (scapula)
   X (sternum)
   X (temporal)
   P (tibia)
   P (ulna)
   X (zygomatic)
4. b
5. d
6. T occipital bone
      foramen magnum
   T sphenoid bone
      optic foramen
   T os coxa
      obturator foramen
   F ethmoid bone
      inferior nasal conchae
   F scapula
      acetabulum
   F humerus
      coronoid process
   F os coxa
      glenoid fossa

   T ulna
      styloid process
7. B (ball and socket joints)
   B (bursae)
   B (hinge joints)
   C (immovable joints)
   A (intervertebral joints)
   C (sutures)
   A (symphysis pubis)
   B (synovial membrane)

☞ **Answers to Terminology Exercises**

Osteogenesis
Acetabulum
Coracoid
Kyphosis
Osteoclast

Like a sieve
Like or resembling a tooth
Pouch or sac
To form bone
Resembling a wedge

D (condition of making blood)
C (like a sieve)
A (presence of little attachments)
E (condition of a "together" joint)
B (pertaining to a little joint)

☞ **Answers to Fun and Games**

```
S • V • • M A S T O I D • S T E M P O R A L
• U • E • • N A S A L • F M U I H C S I • •
• • T • R • • • I F R • O • B I L I U M • •
C • C A • T • • N I • S R • E • D • • • • •
I • L • E A E • F B S L A S R A T A T E M •
A • A P L M L B E U • • M L O • I • R • • •
A • V • H A Y L R L • • E S S B P U B I S •
M • I • C A M R I A M A N D I B L E • L • •
O U C • • A L I O X E • M T T X • • A C • •
G H L • C • R A R T A • A R Y • A P O • P •
Y U E U • A L P N C I M G O • • R C • A • •
Z M M • B U L S A G A D N C • A C • L M • •
• E L U P A A C S L E L U H C Y • A U • S •
• R • A I L T • A • S S M A X • T R • P C •
• U C • T R F E L N • • T N L I C • H M O •
• S P A • I B R C • E E • T N A • E • U N •
• • A F • • P U O A M U • E S • N • • N D •
• • T • E • • I N N X • S R • O • R • R Y •
• • E • • M • L C A T O • • I • R • E E L •
• • L T A L U S H C M A C D D I O M H T E •
• • L A T E I R A P O • L S B • O • • S X •
• • A X I P H O I D • • • • S O V • • • • • E
```

Word List:
Acetabulum
Atlas
Axis
Calcaneus
Carpals
Clavicle

Coccyx
Condyle
Ethmoid
External auditory meatus
Femur
Fibula
Foramen magnum
Frontal
Humerus
Ilium
Inferior nasal concha
Ischium
Lacrimal
Mandible
Manubrium
Mastoid
Maxilla
Metacarpals
Metatarsals
Nasal
Occipital
Os coxa
Palatine
Parietal
Patella
Phalanges
Pubis
Radius
Ribs
Sacrum
Scapula
Sphenoid
Sternum
Talus
Tarsals
Temporal
Tibia
Trochanter
Tuberosity
Ulna
Vertebrae
Vomer
Xiphoid
Zygomatic

☞ **Quiz/Test Questions**

Note: There are fifty multiple-choice questions for this chapter in the computerized test bank.

Name the following:

1. Enlarged ends of long bones.
   Answer: epiphysis.

2. Connective tissue covering around the outside of the diaphysis of a long bone.
   Answer: periosteum.

3. Cell that forms osseous tissue.
   Answer: osteoblast.

4. Cell that is responsible for dissolving bone to enlarge the marrow cavity.
   Answer: osteoclast.

5. Type of bone development process that forms the flat bones of the skull.
   Answer: intramembranous ossification.

6. Region where a bone increases in length.
   Answer: epiphyseal plate.

7. Bone of the face that contains a paranasal sinus.
   Answer: maxilla.

8. Cranial bone that has wings, paranasal sinus, and place for the pituitary gland.
   Answer: sphenoid.

9. Type of vertebrae with articular facets for the ribs.
   Answer: thoracic vertebrae.

10. Top portion of the sternum.
    Answer: manubrium.

11. Anterior bone of the pectoral girdle.
    Answer: clavicle.

12. Bone on medial side of forearm.
    Answer: ulna.

13. Large depression in os coxa for articulation with the femur.
    Answer: acetabulum.

14. Bone that articulates with the distal tibia.
    Answer: talus.

15. Freely movable joints.
    Answer: diarthroses or synarthroses.

True/False Questions:

1. Calcium is stored in the medullary cavity of long bones.
   Answer: False; it is stored in the intercellular matrix of the osteon.

2. Phalanges are short bones.
   Answer: False; they are long bones because their longitudinal dimension is greater than the horizontal dimension.

3. Most of the bones of the body are initially formed as hyaline cartilage models that later develop into bone.
   Answer: True.

4. Osteoporosis often results in an exaggerated thoracic curvature of the vertebral column, which is called kyphosis.
   Answer: True.

Questions for this chapter continue on page 137.

7 Muscular System

Actin Contractile protein in the thin filaments of skeletal muscle cells.

All-or-none principle An individual muscle fiber (cell) contracts to the greatest of its ability or not at all.

Antagonist A muscle that has an action that is opposite to the prime mover.

Aponeurosis Broad, flat sheet of connective tissue that connects one muscle to another or a muscle to bone.

Endomysium Connective tissue that surrounds individual muscle fibers (cells).

Epimysium Fibrous connective tissue that surrounds a whole muscle.

Fasciculus A small bundle or cluster of muscle or nerve fibers (cells); also called **fascicle** plural, **fasciculi.**

Insertion The end of a muscle that is attached to a relatively movable part; the end opposite the origin.

Motor unit A single neuron and all the muscle fibers it stimulates.

Myofibrils Theadlike structures that run longitudinally through muscle cells and composed of actin and myosin myofilaments.

Myofilaments Ultramicroscopic, threadlike structures in the myofibrils of muscle cells; composed of the contractile proteins actin and myosin.

Myosin Contractile protein in the thick filaments of skeletal muscle cells.

Neuromuscular junction The area of communication between the axon terminal of a motor neuron and the sarcolemma of a muscle fiber; also called a myoneural junction.

Neurotransmitter A chemical substance that is released at the axon terminals to stimulate a muscle fiber contraction or an impulse in another neuron.

Origin The end of a muscle that is attached to a relatively immovable part; the end opposite the insertion.

Perimysium Fibrous connective tissue that surrounds a bundle, or fasciculus, of muscle fibers (cells).

Prime mover The muscle that is mainly responsible for a particular body movement; also called agonist.

Sarcolemma The cell membrane of a muscle fiber (cell).

Sarcomere A functional contractile unit in a skeletal muscle fiber.

Sarcoplasm Cytoplasm of muscle fibers (cells).

Sarcoplasmic reticulum Network of tubules and sacs in muscle cells; similar to endoplasmic reticulum in other cells.

Synergist A muscle that assists a prime mover but is not capable of producing the movement by itself.

T tubules Invaginations of the sarcolemma that form transverse tubules in a muscle cell and permit electrical impulses to travel deeper into the cell.

Tendon Band of dense fibrous connective tissue that attaches muscle to bone

☞ **Chapter Objectives**

Upon completion of this chapter the student should be able to:

1. Compare skeletal muscle tissue with visceral and cardiac muscle tissue.
2. List four characteristics of muscle tissue that relate to its functions.
3. List three functions of skeletal muscle tissue.
4. Describe the structure of a skeletal muscle, including its connective tissue coverings.
5. Define the terms origin and insertion as they relate to muscle function.
6. Distinguish between actin and myosin.
7. Identify the bands and lines on the myofibers that make up the striations of skeletal muscle fibers.
8. Define the term sarcomere.
9. Explain why skeletal muscle needs an abundant nerve and blood supply.
10. Describe the sequence of events at the neuromuscular junction that provide the stimulus for contraction of a sarcomere.
11. Describe the changes that take place in actin and myosin myofilaments during contraction.
12. Use the terms threshold and subthreshold stimulus to explain what is meant by the all-or-none principle as it pertains to skeletal muscle contraction.

13. Define the terms motor unit, multiple motor unit summation, twitch, tetanus, multiple wave summation, treppe, and muscle tone.
14. Distinguish between isometric and isotonic contractions.
15. Describe how energy is provided for muscle contraction and how oxygen debt occurs.
16. Distinguish between synergists and antagonists.
17. Describe and illustrate flexion, extension, hyperextension, dorsiflexion, plantar flexion, abduction, adduction, rotation, supination, pronation, circumduction, inversion, and eversion.
18. Describe seven features of a muscle that are frequently used to name the muscle.
19. Locate, identify, and describe the actions of the major muscles of the head and neck.
20. Locate, identify, and describe the actions of the major muscles of the trunk.
21. Locate, identify, and describe the actions of the major muscles of the upper extremity.
22. Locate, identify, and describe the actions of the major muscles of the lower extremity.

☞ **Chapter Outline/Summary**

Characteristics and Functions of the Muscular System (Objectives 1-3)
Comparison of skeletal muscle tissue with other muscle tissues (Objective 1)
- Skeletal muscle cells are cylindrical in shape, with many nuclei, and have striations.
- The muscles are attached to bones and produce movement under voluntary control.

Four characteristics of muscle tissue (Objective 2)
- Excitability is the ability to respond to a stimulus.
- Contractility is the ability to shorten.
- Extensibility is the ability to be stretched.
- Elasticity is the ability to return to normal shape after contraction or extension.

Three functions of skeletal muscle tissue (Objective 3)
- All muscle tissue produces movement. Skeletal muscles produce the movements of the arms and legs, movements involved in facial expressions, and movements that result in breathing.
- Posture is the result of skeletal muscle contraction.
- Muscle contraction produces heat necessary to maintain body temperature.

Structure of Skeletal Muscle (Objectives 4-9)
Whole skeletal muscle (Objective 4)
- A muscle consists of many muscle fibers or cells.
- Each muscle is surrounded by connective tissue epimysium.
- Inward extension of connective tissue separate the muscle into bundles of fibers. Each bundle, called a fasciculus, is surrounded by perimysium.
- Each individual muscle fiber is surrounded by endomysium.

Skeletal muscle attachments (Objective 5)
- Muscles are attached to bones by tendons.
- A broad flat sheet of tendon is an aponeurosis.
- The origin is the less movable end of attachment.
- The insertion is the more movable end.

Skeletal muscle fibers (Objectives 6-8)
- A skeletal muscle fiber is a muscle cell with typical cellular organelles. The cell membrane is the sarcolemma, the cytoplasm is sarcoplasm, and the endoplasmic reticulum is called sarcoplasmic reticulum.
- The sarcoplasm contains myofibrils that are made of still smaller units called myofilaments. Thick myofilaments are composed of the protein myosin and thin myofilaments are formed from the protein actin.
- The characteristic striations of skeletal muscle fibers (the A, I, and H bands) are due to the arrangement of the myosin and actin myofilaments. The Z and M lines are points of myofilament connections.
- A sarcomere is the functional, or contractile, unit of a myofibril. It extends from one Z line to the next Z line.

Nerve and blood supply (Objective 9)
- Muscles must be stimulated by a nerve before they can contract, therefore muscles have an abundant nerve supply.
- An abundant blood supply is necessary to deliver the nutrients and oxygen required for contraction.

Contraction of Skeletal Muscle (Objectives 10-17)
Stimulus for contraction (Objective 10)
- Skeletal muscles are generally stimulated to contract by motor neurons.
- A neuromuscular junction is the region where an axon terminal of a motor neuron is closely associated with a muscle fiber.

- Acetylcholine, a neurotransmitter released by the motor neuron, diffuses across the synaptic cleft to stimulate the sarcolemma.
- Acetylcholinesterase is an enzyme that inactivates acetylcholine to prevent continued contraction from a single impulse.

Sarcomere contraction (Objectives 11 and 12)

- When a muscle fiber is stimulated, an impulse travels down the sarcolemma into a T-tubule, which releases calcium ions from the sarcoplasmic reticulum.
- Calcium alters the configuration of actin and energizes myosin breaking down ATP.
- Cross bridges form between the actin and myosin and energy from ATP results in a power stroke that pulls actin toward the center of the myosin myofilaments. This shortens the length of the sarcomere.
- When stimulation at the neuromuscular junction stops, calcium returns to the sarcoplasmic reticulum, and actin and myosin resume their non-contracting positions.
- The minimum stimulus necessary to cause muscle fiber contraction is a threshold, or liminal, stimulus. A lesser stimulus is sub-threshold, or subliminal.
- Individual muscle fibers contract according to the all-or-none principle. If a threshold stimulus is applied, the fiber will contract; a greater stimulus does not create a more forceful contraction. If the stimulus is sub-threshold, the fiber does not contract at all.

Contraction of a whole muscle (Objectives 13 and 14)

- Whole muscles show varying strengths of contraction due to motor unit and wave summation.
- A motor unit is a single neuron and all the muscle fibers it stimulates.
- Contraction strength of a whole muscle can be increased by multiple motor unit summation, stimulating more motor units.
- A muscle twitch, the response to a single stimulus, shows a lag phase, a contraction phase, and a relaxation phase.
- If a second stimulus of the same intensity as the first is applied before the relaxation phase is complete, a second contraction, stronger than the first results. Repeated stimulation of the same strength that results in stronger contractions is called multiple wave summation.
- Rapid repeated stimulation that allows no relaxation results in a smooth sustained contraction that is stronger than the contraction from a single stimulus of the same intensity. This is tetanus, a form of multiple wave summation. This is the usual form of muscle contraction.
- Treppe is the staircase effect that is evidenced when repeated stimuli of the same strength produce successively stronger contractions. This is due to changes in the cellular environment.
- Muscle tone refers to the continued state of partial contraction in muscles. It is important in maintaining posture and body temperature.
- Isotonic contractions produce movement but muscle tension remains constant. Isometric contractions increase muscle tension but do not produce movement. Most body movements involve both types of contractions.

Energy source and oxygen debt (Objective 15)

- In muscle contraction, energy is needed for the power stroke, detachment of myosin heads, and active transport of calcium.
- ATP provides the initial energy for muscle contraction.
- The ATP supply is replenished by creatine phosphate.
- When muscles are actively contracting for extended periods of time, glucose becomes the primary energy source to produce more ATP.
- When adequate oxygen is available, glucose is metabolized by aerobic respiration to produce ATP. If adequate oxygen is not available, the mechanism for producing ATP from glucose is anaerobic respiration.
- Aerobic respiration produces nearly 20 times more ATP per glucose than the anaerobic pathway, but anaerobic respiration occurs at a faster rate.
- Oxygen debt occurs when there is an accumulation of lactic acid from anaerobic respiration. Additional oxygen from continued labored breathing is needed to repay the debt and restore resting conditions.

Movements (Objectives 16 and 17)

- A prime mover, or agonist, has the major role in producing a specific movement.
- A synergist assists the prime mover.
- An antagonist opposes a particular movement.
- Descriptive terms used to depict particular movements include flexion, extension, hyperextension, dorsiflexion, plantar flexion, abduction, adduction, rotation, supination, pronation, circumduction, inversion, and eversion.

Skeletal Muscle Groups (Objectives 18-22)
Naming muscles (Objective 18)

- Many skeletal muscles have names that describe some feature of the muscle.
- Features used to name muscles include size, shape, direction of the fibers, location, number of attachments, origin, insertion, and action.

Muscle of the head and neck (Objective 19)

- Muscles of facial expression include the frontalis, orbicularis oris, orbicularis oculi, buccinator, and zygomaticus.
- Chewing muscles are called muscles of mastication and include the masseter and temporalis.
- Neck muscles include the sternocleidomastoid, which flexes the neck, and the trapezius, which extends the neck.

Muscles of the trunk (Objective 20)

- Muscles that act on the vertebral column include the erector spinae group and the deep back muscles. The erector spinae form a large muscle mass that extends from the sacrum to the skull on each side of the vertebral column. The deep back muscles are short muscles that occupy the space between the spinous and transverse processes of adjacent vertebrae.
- Muscles of the thoracic wall are involved in the process of breathing. These include the intercostal muscles and the diaphragm.
- Abdominal wall muscles include the external oblique, internal oblique, transversus abdominis, and rectus abdominis. These muscles provide strength and support to the abdominal wall.
- Pelvic floor muscles form a covering for the pelvic outlet and provide support for the pelvic viscera. The superficial muscles form the urogenital diaphragm and the deeper muscles, the levator ani, form the pelvic diaphragm.

Muscles of the Upper Extremity (Objective 21)

- Muscles that move the shoulder and arm
  - The trapezius and serratus anterior muscles attach the scapula to the axial skeleton and move the scapula.
  - The pectoralis major, latissimus dorsi, deltoid, and rotator cuff muscles insert on the humerus and move the arm.
- Muscles that move the forearm and hand
  - The triceps brachii, in the posterior compartment of the arm, extends the forearm.
  - The biceps brachii and brachialis muscles, in the anterior compartment, flex the forearm.
  - The brachioradialis muscle, primarily located in the forearm, flexes and supinates the forearm.
  - Most of the muscles that are located on the forearm act on the wrist, hand, or fingers.

Muscles of the Lower Extremity (Objective 22)

- Muscles that move the thigh
  - The gluteus maximus, gluteus medius, gluteus minimus, and tensor fasciae latae muscles abduct the thigh.
  - The iliopsoas, an anterior muscle, flexes the thigh.
  - The muscles in the medial compartment include the adductor longus, adductor brevis, adductor magnus, and gracilis muscles. These muscles adduct the thigh.
- Muscles that move the leg
  - The quadriceps femoris muscle group, located in the anterior compartment of the thigh, straightens the leg at the knee.
  - The sartorius muscle, a long strap-like muscle on the anterior surface of the thigh, flexes and medially rotates the leg.
  - The hamstring muscles, in the posterior compartment of the thigh, are antagonists to the quadriceps femoris muscle group.
- Muscles that move the ankle and foot
  - The principle muscle in the anterior compartment of the leg is the tibialis anterior, which dorsiflexes the foot.
  - Contraction of the peroneus muscles, in the lateral compartment of the leg, everts the foot.
  - The gastrocnemius and soleus form the bulky mass of the posterior compartment of the leg. These muscles plantar flex the foot.

☞ **Answers to Review Questions**

1. Skeletal muscle cells are cylindrical, have many peripherally located nuclei, and are striated. Visceral muscle cells are spindle-shaped with tapered ends, have a single centrally located nucleus and do not have striations.
2. Excitability, contractility, extensibility, and elasticity.
3. Maintain posture, lend stability to joints, and heat production.
4. Endomysium is the connective tissue covering around each individual muscle fiber. Perimysium is the connective tissue covering that encloses a bundle (fasciculus) of muscle

fibers. Epimysium is the connective tissue covering around an entire muscle.

5. Insertion

6. Thick myofilaments: myosin
   Thin myofilaments: actin

7. A band: myosin with actin overlapping at the ends
   I band: actin only
   H band: myosin only

8. Sarcomere

9. Skeletal needs an abundant blood supply to bring oxygen to the contracting muscle and to carry away the carbon dioxide. Contracting muscle requires nutrients which are brought by the blood. The nerve supply is needed because muscle fibers have to receive a stimulus from nerves before they contract.

10. Acetylcholine

11. In a resting fiber, troponin covers the myosin binding sites. In response to a stimulus, calcium ions react with troponin, which changes the configuration, and exposes the binding sites so that myosin can form cross bridges with the actin. Calcium is responsible for the changes.

12. In response to a threshold stimulus, a muscle fiber contracts to its maximum ability under the existing conditions, while a lesser stimulus elicits no contraction.

13. A motor unit is a single nerve fiber and all the muscle fibers it stimulates.

14. In response to a stronger stimulus, additional motor units contract to produce a stronger contraction in the whole muscle. This is multiple motor unit summation. Rapid repeated stimulation produces contractions that merge into a smooth, sustained contraction that is stronger than the contraction from a single stimulus of the same intensity. This is multiple wave summation.

15. Isotonic

16. Immediate source is ATP.
    Intermediate source is creatine phosphate which is used to replenish the ATP.
    Long term sources are glucose and fatty acids that are metabolized to produce energy.

17. Lactic acid

18. A synergist facilitates, or assists, the action of a prime mover. An antagonist has the opposite action.

19. Extension is opposite of flexion.
    Adduction is opposite of abduction.
    Supination is opposite of pronation.
    Plantar flexion is opposite of dorsiflexion.

20. **Size**: Vastus lateralis, vastus medialis, vastus intermedius, gluteus maximus, adductor magnus, adductor longus, adductor brevis, gluteus minimus.
    Shape: Deltoid, rhomboideus, latissimus dorsi, teres major, teres minor, trapezius.
    Location: Pectoralis major and minor, gluteus maximus, gluteus minimus, gluteus medius, biceps brachii, triceps brachii, supraspinatus, infraspinatus, subscapularis, vastus lateralis.
    Number of origins: Triceps brachii, biceps brachii, quadriceps femoris.

21. (a) Orbicularis oris and orbicularis oculi
    (b) Zygomaticus
    (c) Temporalis and masseter
    (d) Trapezius

22. External oblique, internal oblique, transversus abdominis, rectus abdominis

23. Internal intercostals

24. Trapezius and serratus anterior

25. (a) Abduct the arm
    (b) Adduct and flex the arm
    (c) Adduct and medially rotate arm

26. (a) Brachialis and brachioradialis
    (b) Triceps brachii

27. (a) Tensor fasciae latae
    (b) Gluteus maximus

28. Adductor longus, adductor brevis, adductor magnus, gracilis

29. Biceps femoris, semimembranosus, and semitendinosus flex the leg at the knee.

30. Vastus lateralis, vastus medialis, vastus intermedius, and rectus femoris extend the leg at the knee.

31. Sartorius

32. Gastrocnemius and soleus

33. Tibialis anterior

☞ **Answers to Learning Exercises**

Characteristics and Functions of the Muscular System (Objectives 1-3)

1. C (striated and involuntary)
   B (spindle shaped fibers)
   A (multinucleated and cylindrical)
   B (found in blood vessels)
   C (found in the heart)

2. Excitability, contractility, extensibility, elasticity

3. Movement, posture, heat production

Structure of Skeletal Muscle (Objectives 4-9))

1. Insertion
   Endomysium
   Myosin
   Fasciculus
   Origin
   Aponeurosis

Sarcolemma
Actin
Sarcomere
Thin/actin

2. Muscle fibers must be stimulated before they can contract, therefore they have an abundant **nerve supply**. The blood supply delivers **oxygen** and **nutrients** for contraction.

Contraction of Skeletal Muscle (Objectives 10-17)

1. 7 Energized myosin
   2 Acetylcholine is released
   8 Power stroke pulls
   6 Calcium reacts
   3 Acetylcholine reacts
   1 Nerve impulse
   5 Calcium ions are released
   4 Impulse travels
2. Threshold (liminal)
   All-or-none
   Subthreshold (subliminal)
   Motor unit
   Summation
   Tetany
   Treppe
   Tone
   Isotonic
   Isometric
3. ATP
   Creatine phosphate
   Myoglobin
   Lactic acid
   $CO_2 + H_2O$ + energy (ATP)
4. The muscle that has a primary role in providing a movement is called the **prime mover**. Muscles that assist this muscle are called **synergists** and muscles that oppose the movement are called **antagonists**.
5. E (extension of foot to stand on tiptoes)
   G (movement of arm toward the midline)
   B (straightening the arm at the elbow)
   I (turning the palm of the hand forward)
   H (turning the head from side to side)
   A (moving elbow to put hand on shoulder)
   L (turning the sole of foot inward)
   F (spreading the fingers apart)
   C (tilting head backward)
   K (drawing circles on chalkboard)

Skeletal Muscle Groups (Objectives 18-22)

1. Action (adductor)
   Direction of fibers (rectus)
   Size (maximus)
   Location (gluteus)
   Shape (deltoid)
   Number of origins (biceps)
   Origin/insertion (Brachioradialis)
   Location (pectoralis)
2. Frontalis
   Orbicularis oris
   Temporalis
   Masseter
   Orbicularis oculi
   Sternocleidomastoid
   Diaphragm
   Pectoralis major
   Latissimus dorsi
   Deltoid
   Triceps brachii
   Biceps brachii
   Brachialis
   Iliopsoas
   Tensor fasciae latae
   Gluteus maximus
   Adductors
   Quadriceps femoris
   Biceps femoris
   Semimembranosus
   Semitendinosus
   Gastrocnemius
   Soleus (also peroneus)
   Tibialis anterior
3. A Frontalis
   B Orbicularis oculi
   C Orbicularis oris
   D Temporalis
   E Zygomaticus
   F Buccinator
   G Trapezius
   H Infraspinatus
   I Latissimus dorsi
   J Gluteus maximus
   K Sternocleidomastoid
   L Deltoid
   M Triceps brachii
   N Gluteus medius
4. A Iliopsoas
   B Adductor longus
   C Gracilis
   D Vastus medialis
   E Tensor fasciae latae
   F Sartorius
   G Vastus medialis
   H Vastus lateralis
   I Gluteus medius
   J Vastus lateralis
   K Semitendinosus
   L Biceps femoris
   M Gluteus maximus
   N Gracilis
   O Semimembranosus
   P Gastrocnemius

☞ Answers to Chapter Self-Quiz

1. X before responds to a stimulus, multinucleated, and striations.
2. Perimysium
3. Origin
4. b (I band)
5. Sarcolemma
   Acetylcholine
   Myosin
   Lactic acid
   Sarcoplasmic reticulum
   ATP
   Isotonic
   Twitch
   Tetany
   Treppe
6. b (rotation - pronation)
7. d (flexion)
8. F (closes jaw)
   P (elevates corner of mouth)
   N (extends the head)
   I (long straight muscle)
   K (moves the scapula)
   G (adducts the arm)
   C (antagonist of latissimus dorsi)
   B (inserts on radius)
   O (antagonist of biceps brachii)
   A (located in medial thigh)
   L (most superficial and lateral)
   J (long straplike muscle)
   H (muscle group that extends leg)
   D (located in posterior compartment)
   E (muscle group that flexes leg)
   M (muscle that allows to "walk on heels")

☞ Answers to Terminology Exercises

Actin
Diaphragm
Isometric
Masseter
Abduct

Triangle shape
Two heads
Same tension
Bending
Covering of muscle cell

B (neutral substance in muscle)
D (work together)
C (muscle matter)
A (without development, wasting)
E (stiff)

☞ Answers to Fun and Games

1. Contractility (20 points)
   Extensibility (21 points)
   Excitability (21 points)
   Elasticity (14 points)
2. Endomysium (15 points)
   Epimysium (15 points)
   Perimysium (16 points)
3. Actin (6 points)
   Myosin (9 points)
4. Origin (8 points)
   Insertion (9 points)
5. Lag (5 points)
   Contraction (13 points)
   Relaxation (15 points)
6. Flexion (14 points)
   Extension (13 points)
   Abduction (12 points)
   Adduction (12 points)
   Supination (12 points)
   Pronation (11 points)
   Circumduction (18 points)
   Rotation (8 points)
   Inversion (12 points)
   Eversion (11 points)
   Dorsiflexion (20 points)
7. Rectus (6 points)
   Orbicularis (14 points)
   Transversus (14 points)
   Oblique (13 points)
8. Frontalis (12 points)
   Zygomaticus (21 points)
   Buccinator (13 points)
9. Deltoid (10 points)
10. Gastrocnemius (17 points)
    Soleus (7 points)

☞ Quiz/Test Questions

Note: There are fifty multiple-choice questions for this chapter in the computerized test bank.

Name the following:

1. A function of skeletal muscle tissue in addition to movement and posture.
   Answer: heat production.

2. Broad flat sheet of tendon.
   Answer: aponeurosis.

3. Protein in thick myofilaments..
   Answer: myosin.

4. Functional unit of a muscle fiber, from Z line to Z line.
   Answer: sarcomere.

Questions for this chapter continue on page 137.

8 Nervous System

☞ **Key Terms/Concepts**

Action potential A nerve impulse; a rapid change in membrane potential that involves depolarization and repolarization.

Adrenergic fiber A nerve fiber that releases epinephrine (adrenaline) or norepinephrine (noradrenalin) at a synapse.

Afferent neuron Nerve cell that carries impulses toward the central nervous system from the periphery; sensory neuron.

Association neuron Nerve cell, totally within the central nervous system, that carries impulses from a sensory neuron to a motor neuron; also called interneuron.

Autonomic nervous system the portion of the peripheral efferent nervous system consisting of motor neurons that control involuntary actions.

Axon The single efferent process of a neuron that carries impulses away from the cell body.

Brain stem The portion of the brain, between the diencephalon and spinal cord, that contains the midbrain, pons, and medulla oblongata.

Central nervous system (CNS) The portion of the nervous system that consists of the brain and spinal cord.

Cerebellum Second largest part of the human brain, located posterior to the pons and medulla oblongata, and involved in the coordination of muscular movements.

Cerebrospinal fluid A fluid, similar to plasma that fills the subarachnoid space around the brain and spinal cord and is in the ventricles of the brain.

Cerebrum The largest and uppermost part of the human brain, which is concerned with consciousness, learning, memory, sensations, and voluntary movements.

Cholinergic fiber A nerve fiber that releases acetylcholine at a synapse.

Dendrite The branching afferent process of a neuron that receives impulses from other neurons and transmits them toward the cell body.

Diencephalon Parts of the brain between the cerebral hemispheres and the midbrain; includes the thalamus, hypothalamus, and epithalamus.

Efferent neuron Nerve cell that carries impulses away from the central nervous system toward the periphery; motor neuron.

Ganglion A group of nerve cell bodies that lie outside the central nervous system; plural ganglia.

Medulla oblongata Lowest part of the brain stem, which contains the vital cardiac center, vasomotor center, and respiratory center.

Membrane potential The difference in electrical charge between inside and outside a cell membrane.

Meninges Connective tissue membranes that surround the brain and spinal cord.

Midbrain Region of the brainstem between the diencephalon and the pons.

Myelin White, fatty substance that surrounds many nerve fibers.

Neurilemma The layer of Schwann cells that surrounds a nerve fiber in the peripheral nervous system and, in some cases, produces myelin; also called neurolemma or Schwann's sheath.

Neuroglia Supporting cells of the nervous system that do not conduct impulses.

Neuron A nerve cell, including its processes.

Neuronal pool Functional group of neurons within the central nervous system that receives information, processes and integrates that information, then transmits it to some other destination.

Neurotransmitter A chemical substance that is released at the axon terminals to stimulate impulse conduction in another neuron.

Node of Ranvier Short space between two segments of myelin in a myelinated nerve fiber.

Parasympathetic division One of two divisions of the autonomic nervous system; primarily concerned with processes that involve the conservation and restoration of energy; sometimes referred to as the "rest and repose" division; also called the craniosacral division.

Peripheral nervous system (CNS) The portion of the nervous system that is outside of the brain and spinal cord; consists of the nerves and ganglia.

Pons Middle portion of the brain stem, between the midbrain and the medulla oblongata.

Refractory period Time during which an excitable cell cannot respond to a stimulus that is usually adequate to start an action potential.

Reflex arc Smallest unit of the nervous system that can receive a stimulus and generate a response.

Saltatory conduction Process in which a nerve impulses travels along a myelinated nerve fiber by jumping from one node of Ranvier to the next.

Somatic nervous system The portion of the peripheral efferent nervous system consisting of motor neurons that control voluntary actions of skeletal muscles.

Spinal cord Portion of the central nervous system that extends inferiorly from the brain and provides conduction to and from the brain.

Sympathetic division One of the two divisions of the autonomic nervous system; primarily concerned with processes that involve the expenditure of energy; sometimes referred to as the "fight or flight" division; also called the thoracolumbar division.

Synapse The region of communication between two neurons.

Threshold stimulus Minimum level of stimulation that is required to start a nerve impulse.

☞ Chapter Objectives

Upon completion of this chapter the student should be able to:

1. Describe three types of nervous system functions.
2. Outline the organization of the nervous system.
3. Identify the two categories of cells in nerve tissue.
4. Describe the structure of a neuron.
5. Identify the parts of a neuron on a diagram or model.
6. Classify neurons according to their function.
7. Name, identify the location, and state the functions of six types of neuroglia cells.
8. State two functional characteristics of neurons.
9. Describe the resting polarity of the resting cell membrane of a neuron.
10. Describe the sequence of events that lead to an action potential when the cell membrane is stimulated.
11. Explain how the cell membrane is restored to resting conditions after an action potential.
12. Define the terms threshold stimulus and subthreshold stimulus as they pertain to action potentials.
13. Explain how an impulse is conducted along the length of a neuron.
14. Define the terms saltatory conduction, refractory period, and all-or-none principle as they pertain to nerve impulses.
15. Describe the structure of a synapse and how an impulse is conducted from one neuron to another across the synapse.
16. Distinguish between excitatory transmission and inhibitory transmission.
17. Describe three types of circuits in neuronal pools.
18. List the five basic components of a reflex arc.
19. Describe the three layers of meninges around the central nervous system.
20. Locate and identify the lobes of the cerebral hemispheres, gray matter, white matter, and basal ganglia.
21. Locate the primary motor, sensory, and association areas of the cerebrum.
22. Describe the location, structure, and functions of the diencephalon.
23. Identify the three regions of the brainstem and describe the functions of each region.
24. Describe the structure and functions of the cerebellum.
25. Trace the flow of cerebrospinal fluid from its origin in the ventricles to its return to the blood.
26. Describe the structure and functions of the spinal cord.
27. Use the terms endoneurium, perineurium, epineurium, and fasciculus to describe the structure of a nerve.
28. List the twelve cranial nerves and state the function of each one.
29. Describe the composition of spinal nerves, list the five groups of spinal nerves, and state the number of nerves in each group.
30. Define the term plexus and identify at least one nerve that arises from each of the cervical, brachial, and lumbosacral plexuses.
31. Compare the structural and functional differences between the somatic efferent pathways and the autonomic nervous system.
32. Distinguish between the sympathetic and parasympathetic divisions of the autonomic nervous system in terms of structure, function, and neurotransmitters.

☞ **Chapter Outline/Summary**

Functions of the Nervous System (Objective 1)

- The activities of the nervous system can be grouped into sensory, integrative, and motor functions.

Organization of Nervous System (Objective 2)

- The nervous system is divided into the central nervous system and the peripheral nervous system.
- The peripheral nervous system consists of an afferent (sensory) division and an efferent (motor) division.
- The efferent (motor) division of the peripheral nervous system is divided into the somatic nervous system and the autonomic nervous system.

Nerve Tissue (Objectives 3-7)

Neurons (Objectives 3-6)

- Neurons are the nerve cells that transmit impulses. They are amitotic.
- The three components of a neuron are a cell body or soma, one or more dendrites, and a single axon.
- Many neurons are surrounded by segmented myelin. The gaps in the myelin are the nodes of Ranvier. An outer covering, the neurilemma, plays a role in nerve regeneration.
- Axons terminate in telodendria, which have synaptic bulbs on the distal end.
- Functionally, neurons are classified as afferent (sensory), efferent (motor), or interneurons (association).

Neuroglia (Objectives 3 and 7)

- Neuroglia support, protect, and nourish the neurons. They are capable of mitosis.
- Neuroglia cells include astrocytes, microglia, ependymal, oligodendrocytes, Schwann cells and satellite cells.

Nerve Impulses (Objectives 8-18)

Resting membrane (Objectives 8 and 9)

- Excitability, the ability to respond to a stimulus, and conductivity, the ability to transmit an impulse, are two functional characteristics of neurons.
- The cell membrane of a non-conducting neuron is polarized with an abundance of sodium ions outside the cell and an abundance of potassium and negatively charged proteins inside the cell. The inside of the membrane is approximately 70 millivolts negative to the outside.

Stimulation of a neuron (Objectives 10-12)

- In response to a stimulus, the cell membrane becomes permeable to sodium ions, so they rapidly enter the cell and depolarize the membrane.
- Continued sodium ion diffusion causes reverse polarization.
- After reverse polarization, the membrane becomes impermeable to sodium and permeable to potassium. Potassium diffuse out of the cell to repolarize the membrane.
- The action potential is a result of depolarization, reverse polarization, and repolarization of the cell membrane.
- After the action potential, active transport mechanisms move sodium out the cell and potassium into the cell to restore resting conditions.
- A threshold stimulus is the minimum stimulus necessary to start an action potential. A weaker stimulus is subthreshold.

Conduction along a neuron (Objectives 13 and 14)

- An action potential at a given point stimulates depolarization at an adjacent point to create a propagated action potential, or nerve impulse, that continues along the entire length of the neuron.
- Saltatory conduction occurs in myelinated fibers where the action potential "jumps" from node to node.
- The refractory period is the time during which the cell membrane is recovering from depolarization. Absolute refractory period is the time during which the membrane is permeable to sodium ions and cannot respond to a second stimulus so matter how strong. The relative refractory period is roughly comparable to the time when the membrane is permeable to potassium. During this time, it takes a stronger than normal stimulus to initiate an action potential.
- If a threshold stimulus is applied, an action potential is generated and propagated along the entire length of the neuron at maximum strength and speed for the existing conditions. This is the all-or-none principle.

Conduction across a synapse (Objectives 15-17)

- A synapse, the region of communication between two neurons, consists of a synaptic knob, a synaptic cleft, and the postsynaptic membrane.
- At the synapse, a neurotransmitter, such as acetylcholine, diffuses across the synaptic cleft and reacts with receptors on the postsynaptic membrane.
- In excitatory transmission, the reaction between the neurotransmitter and receptor

depolarizes the postsynaptic membrane and initiates an action potential.

- In inhibitory transmission, the reaction between the neurotransmitter and receptor opens the potassium channels and hyperpolarizes the membrane, which makes it more difficult to generate an action potential.
- In a single series circuit, a single neuron synapses with another neuron; in a divergence circuit, a single neuron synapses with multiple neurons; and in a convergence circuit, several presynaptic neurons synapse with a single postsynaptic neuron.

Reflex arcs (Objective 18)

- The reflex arc is a type of conduction pathway and represents the functional unit of the nervous system.
- A reflex arc utilizes a receptor, sensory neuron, center, motor neuron, and effector.

Central Nervous System (Objectives 19-26)

Meninges (Objective 19)

- The brain and spinal cord are covered by three layers of connective tissue membranes called meninges.
- The outer layer of the meninges is the dura mater, middle layer is the arachnoid, and the inner layer is the pia mater.
- The subarachnoid space, between the arachnoid and pia mater, contains blood vessels and cerebrospinal fluid.

Brain (Objectives 20-24)

- Cerebrum (Objectives 20 and 21)
  - The largest portion of the brain is the cerebrum, which is divided by a longitudinal fissure into two cerebral hemispheres. These are connected by a band of white fibers called the corpus callosum. The surface is marked by gyri separated by sulci.
  - Each cerebral hemisphere is divided into frontal, parietal, occipital, and temporal lobes, and an insula.
  - The outer surface of the cerebrum is the cerebral cortex and is composed of gray matter. This surrounds the white matter, which consists of myelinated nerve fibers. Basal ganglia are regions of gray matter scattered throughout the white matter.
  - The primary sensory area is the postcentral gyrus in the parietal lobe; the sensory area for vision is in the occipital lobe; the area for hearing is in the temporal lobe; the olfactory area is in the temporal lobe; and the sensory area for taste is in the parietal lobe.
  - The primary motor area is the precentral gyrus in the frontal lobe.
  - Association areas analyze, interpret, and integrate information. They are scattered throughout the cortex.
- Diencephalon (Objective 22)
  - The diencephalon is nearly surrounded by the cerebral cortex. It includes the thalamus, hypothalamus, and epithalamus.
  - The largest region of the diencephalon is the thalamus, which serves as a relay station for sensory impulses going to the cerebral cortex.
  - The hypothalamus is a small region below the thalamus. It plays a key role in maintaining homeostasis.
  - The epithalamus is the superior portion of the diencephalon and the pineal gland extends from its posterior margin.
- Brainstem (Objective 23)
  - The brainstem is the region between the diencephalon and the spinal cord.
  - The midbrain, the most superior portion of the brainstem, includes the cerebral peduncles and corpora quadrigemina. The midbrain contains voluntary motor tracts, and visual and auditory reflex centers.
  - The pons is the middle portion of the brainstem. It contains the pneumotaxic and apneustic areas that help regulate breathing movements.
  - The medulla oblongata is inferior to the pons and is continuous with the spinal cord. It contains ascending and descending nerve fibers. It also contains the vital cardiac, vasomotor, and respiratory centers.
- Cerebellum (Objective 24)
  - The cerebellum consists of two hemispheres connected by a central region called the vermis. A thin layer of gray matter, the cortex, surrounds the white matter.
  - Cerebellar peduncles connect the cerebellum to other parts of the CNS.
  - The cerebellum is a motor area that coordinates skeletal muscle activity and is important in maintaining muscle tone, posture, and balance.

Ventricles and cerebrospinal fluid (Objective 25)

- Four interconnected cavities, called ventricles, contain cerebrospinal fluid within the brain.

- Cerebrospinal fluid is formed as a filtrate from the blood in the choroid plexus in the ventricles.
- CSF moves from the lateral ventricles, through the interventricular foramen to the third ventricle, through the cerebral aqueduct to the fourth ventricle. From the fourth ventricle, it enters the subarachnoid space and filters through arachnoid granulations into the dural sinuses and is returned to the blood.

Spinal cord (Objective 26)

- The spinal cord begins at the foramen magnum, as a continuation of the medulla oblongata, and extends to the first lumbar vertebra.
- The central core of the spinal cord is gray matter, which is divided into regions called dorsal, lateral, and ventral horns. The white matter, which surrounds the gray matter, is divided into dorsal, lateral, and ventral funiculi, or columns.
- The spinal cord is a conduction pathway and a reflex center.
- Ascending tracts in the spinal cord conduct sensory impulses to the brain. Conduction pathways that carry motor impulses from the brain to effectors are descending tracts.

Peripheral Nervous System (Objectives 27-32)

- The cranial and spinal nerves form the peripheral nervous system. Nerves are classified as sensory, motor, or mixed, depending on the types of fibers they contain.

Structure of a nerve (Objective 27)

- Nerves are bundles of nerve fibers. Each individual nerve fiber is covered by endoneurium. A bundle of nerve fibers, surrounded by perineurium, is called a fasciculus. A nerve contains many fasciculi collected together and surrounded by epineurium.

Cranial nerves (Objective 28)

- Twelve pairs of cranial nerves emerge from the inferior surface of the brain. Cranial nerves are designated by name and by Roman numerals.

Spinal nerves (Objectives 29 and 30)

- All spinal nerves are mixed nerves. They are connected to the spinal cord by dorsal roots, which have only sensory fibers, and ventral roots, which have only motor fibers.
- The 31 pairs of spinal nerves are grouped according to the region of the cord from which they originate. There are 8 cervical nerves, 12 thoracic nerves, 5 lumbar nerves, 5 sacral nerves, and 1 coccygeal nerve.
- In all but the thoracic region, the main portions of the spinal nerves form complex

networks called plexuses. Named nerves exit the plexuses to supply specific regions of the body.

Autonomic nervous system (Objectives 31 and 32)

- The ANS is a visceral efferent system that innervates smooth muscle, cardiac muscle, and glands.
- An autonomic pathway consists of two neurons. A preganglionic neuron leaves the CNS and synapses with a postganglionic neuron in an autonomic ganglion.
- The sympathetic division is also called the thoracolumbar division because its preganglionic neurons originate in the thoracic and lumbar regions of the spinal cord. It is an energy expending system that prepares the body for emergency or stressful conditions.
- The parasympathetic division is also called the craniosacral division because its preganglionic neuron originate in the brain and the sacral region of the spinal cord. It is an energy conserving system that is most active when the body is in a normal relaxed condition.

☞ **Answers to Review Questions**

1. Sensory, integrative, and motor functions
2. Central nervous system (CNS) and peripheral nervous system (PNS). The CNS includes the brain and spinal cord. The PNS includes the cranial nerves and spinal nerves.
3. Neurons and neuroglia
4. Cell body, dendrites, and axon
5. Include in drawing: cell body, nucleus, dendrites, axon, myelin, node of Ranvier, and axon collateral.
6. Afferent or sensory neurons carry impulses toward the CNS. They usually have long dendrites and short axons. Efferent or motor neurons carry impulses away from the CNS. They usually have short dendrites and long axons. Interneurons or association neurons are entirely within the CNS and are located between the afferent and efferent neurons.
7. Neuroglia cells support, protect, and nourish the neurons. There are six types of neuroglia cells.
8. Excitability (irritability) and conductivity.
9. Resting potential is the difference in charges on the two sides of a nonconducting membrane. The inside of the membrane is approximately 70 mV negative to the outside.
10. A stimulus alters the permeability of the cell membrane. First sodium rapidly diffuses into the cell to depolarize the membrane. As

sodium continues to diffuse inward, the inside of the membrane becomes positive to the outside (reverse polarization). Potassium then diffuses out of the cell to repolarize the membrane. An action potential is the rapid sequence of depolarization, reverse polarization, and repolarization.

11. Resting conditions are restored by the active transport sodium/potassium pump.

12. Threshold stimulus is the minimum stimulus necessary to initiate an action potential.

13. A propagated action potential is a nerve impulse that travels along the length of a neuron.

14. Saltatory conduction, which occurs on myelinated axons, is faster than other action potentials because it "jumps" from node to node.

15. Include the synaptic knob with synaptic vesicles of neurotransmitter, the synaptic cleft, and the postsynaptic membrane with binding sites for the neurotransmitter.

16. An excitatory neurotransmitter depolarizes the postsynaptic membrane and initiates an action potential. An inhibitory neurotransmitter hyperpolarizes the postsynaptic membrane, which makes it more difficult to start an action potential.

17. In a convergence circuit, many presynaptic neurons synapse with a single postsynaptic neuron. In a divergence circuit, a single presynaptic neuron synapses with many postsynaptic neurons.

18. Include the receptor, sensory neuron, CNS, motor neuron, and effector.

19. The three layers of meninges are dura mater, arachnoid, and pia mater. They cushion and protect the CNS.

20. Frontal, parietal, temporal, occipital, and insula.

21. Basal ganglia are regions of gray matter that are scattered throughout the white matter of the cerebral hemispheres.

22. a. Postcentral gyrus of parietal lobe
    b. Precentral gyrus of frontal lobe
    c. Occipital lobe
    d. Temporal lobe

23. The diencephalon is centrally located and surrounded by the cerebral hemispheres. It includes the thalamus, hypothalamus, and epithalamus. The thalamus is a relay station for sensory impulses going to the cerebral cortex. The hypothalamus regulates visceral activities, largely through the autonomic nervous system. The epithalamus contains the pineal gland that functions in the onset of puberty and regulates the daily rhythmic cycles in the body.

24. The three parts of the brainstem are the midbrain, pons, and medulla oblongata. The cerebral peduncles of the midbrain contain the voluntary motor tracts that descend from the cerebral cortex. The corpora quadrigemina of the midbrain function in auditory and visual reflexes. The medulla oblongata contains all sensory and motor fibers that go between the brain and spinal cord.

25. Cardiac center, vasomotor center, and respiratory center.

26. The cerebellum is a motor area that coordinates skeletal muscle activity and is important in maintaining muscle tone, posture, and balance.

27. Cerebrospinal fluid is the fluid within the ventricles of the brain and in the subarachnoid space around the brain and spinal cord. It is produced by the choroid plexus within the ventricles, passes into the subarachnoid space, and reenters the blood through arachnoid granulations into the venous sinuses of the dura mater.

28. In the cerebrum, the gray matter is the superficial layer surrounding the white matter. In the spinal cord, the white matter surrounds the centrally located gray matter. The white matter in the cord contains the ascending and descending tracts of myelinated fibers. The gray matter is a region of synapse.

29. Endoneurium around each individual nerve fiber; perineurium around a bundle, or fasciculus, of fibers; epineurium around the entire nerve.

30. Olfactory: sense of smell
    Optic: vision
    Oculomotor: movement of eyes and eyelids
    Trochlear: movement of eyes
    Trigeminal: sensory from face, motor to muscles of mastication
    Abducens: eye movement
    Facial: taste from anterior 2/3 of tongue and motor to muscles of facial expression
    Vestibulocochlear: hearing and equilibrium
    Glossopharyngeal: taste from posterior 1/3 of tongue and motor to pharyngeal muscles and salivary glands
    Vagus: sensory from visceral organs, motor to visceral organs and to pharyngeal muscles

    Accessory: Contraction of trapezius and sternocleidomastoid muscles
    Hypoglossal: contraction of tongue muscles

31. All spinal nerves are mixed nerves that emerge from the spinal cord by a dorsal and ventral root. There are 31 spinal nerves: 8 cervical, 12 thoracic, 5 lumbar, 5 sacral, and 1 coccygeal.

32. A plexus is a complex network of nerve fibers.

33. The pathways in the autonomic nervous system have two neurons while the somatic efferent pathways have only one neuron. In the ANS, the two neurons synapse in ganglia that are outside the CNS.

34. Sympathetic nervous system: thoracolumbar division; neurons synapse in paravertebral and collateral ganglia; have short preganglionic fibers and long postganglionic fibers; show divergence; neurotransmitters are acetylcholine from preganglionic fibers and norepinephrine from postganglionic fibers; energy expending system that helps body deal with emergencies and stressful situations. Parasympathetic nervous system: craniosacral division; neurons synapse in terminal ganglia near the organ innervated; have long preganglionic fibers and short postganglionic fibers; show little or no divergence; neurotransmitter is acetylcholine from both pre and postganglionic fibers; energy conserving system that is most active under ordinary, relaxed conditions.

☞ **Answers to Learning Exercises**

Functions of the Nervous System (Objective 1)
1. The activities of the nervous system are grouped into three functional categories. These are **sensory, integrative, and motor**.

Organization of the Nervous System (Objective 2)
1. The two components of the central nervous system are the **brain** and **spinal cord**.
2. The two divisions of the peripheral nervous system are the **afferent**, or sensory division, and the **efferent**, or motor, division.
3. The autonomic nervous system is a part of the **efferent (motor)** division of the peripheral nervous system.

Nerve Tissue (Objectives 3-7)
1. Dendrite
   Myelin
   Afferent (sensory)
   Node of Ranvier
   Interneurons (association neurons)
   Axon
   Neurilemma
2. A Nucleus
   B Dendrite
   C Schwann cell (neurilemma)
   D Axon collateral
   E Myelin
   F Cell body
   G Node of Ranvier
   H Axon cylinder
   I Telodendria
3. Astrocyte
   Microglia
   Oligodendroglia
   Schwann cell
   Ependyma
   Satellite cell

Nerve Impulses (Objective 8-18)
1. Two functional characteristics of neurons are **irritability (excitability)** and **conductivity**.
2. At rest, the outside of a neuron is **positive** (charge) and has a higher concentration of **sodium** ions relative to the inside. A stimulus changes the permeability of the membrane and it depolarizes. During depolarization, **sodium** ions diffuse into the cell. This creates an **action** potential, or nerve impulse. During repolarization, **potassium** ions diffuse out of the cell. At the conclusion of the **stimulus**, the **sodium/potassium** pump restores the ionic conditions of a resting membrane. The minimum stimulus required to initiate a nerve impulse is called a **threshold (liminal)** stimulus.
3. 11, 9, 6, 3, 13, 7, 4, 1, 12, 8, 5, 2, 10
4. H (rapid conduction)
   I (region of communication)
   E (diffuses across synaptic cleft)
   G (time during which a neuron...)
   C (synaptic transmission...)
   F (functional unit of nervous system)
   B (single neuron synapses...)
   D (synaptic transmission ...)
   A (several neurons synapse...)

Central Nervous System (Objectives 19- 26)
1. 4 (arachnoid)
   2 (dura mater)
   1 (epidural)
   6 (pia mater)
   5 (subarachnoid)
   3 (subdural)
2. A Parietal lobe
   B Occipital lobe
   C Cerebellum
   D Medulla oblongata

E Corpus callosum
F Frontal lobe
G Midbrain
H Pons
3. Cerebellum
   Brainstem
   Diencephalon
   Medulla oblongata
   Hypothalamus
   Pons
   Midbrain
   Cerebellum
   Medulla oblongata
   Pons
4. 7, 2, 6, 9, 4, 1, 3, 8, 5
5. A (central canal)
   B (column of white matter)
   E (dorsal root ganglion)
   D (dorsal horn)
   F (spinal nerve)
   C (ventral root)
6. The spinal cord consists of a central core of white matter surrounded by gray matter.
   The spinal cord consists of a central core of **gray** matter surrounded by **white** matter.

   Two enlargements of the spinal cord are in the thoracic and lumbar regions.
   Two enlargements of the spinal cord are in the **cervical** and **sacral** regions.

   The dorsal, lateral, and ventral horns contain bundles of nerve fibers, called nerve tracts.
   The dorsal, lateral, and ventral **columns** contain bundles of nerve fibers, called nerve tracts.

   Ascending tracts carry motor impulses to the brain.
   Ascending tracts carry **sensory** impulses to the brain.

   Corticospinal tracts are ascending tracts that begin in the cerebral cortex.
   Corticospinal tracts are **descending** tracts that begin in the cerebral cortex.

Peripheral Nervous System (Objectives 27-32)
1. Within a nerve, each individual nerve fiber has a connective tissue covering called **endoneurium**. The nerve itself is covered by connective tissue called **epineurium**.
2. Three nerves that are sensory only: Olfactory, optic, vestibulocochlear

   Nerve to the muscles of facial expression: Facial

Three nerves that function in eye movement: Oculomotor, abducens, trochlear

Nerve that is likely to be involved if you have a toothache in the lower jaw: Mandibular branch of the trigeminal nerve

Nerve to the muscles of mastication: Trigeminal

Nerve that allows you to nod your head: Spinal accessory

3. 31 pr (number of spinal nerves)
   Dorsal (spinal nerve root...)
   Plexus (complex network...)
   8 pr (number of cervical nerves)
   Ventral (spinal nerve...)
   Brachial (nerve plexus that supplies...)
   Cervical (nerve plexus that gives...)
   Lumbosacral (nerve plexus that...)
4. Single arrow red; two arrows blue
   Preganglionic is CNS to ganglion
   Postganglionic is ganglion to effector
   Autonomic effector has asterisk
5. Checks (✓) for Sympathetic
      Arises from thoracic and lumbar regions
      Has short preganglionic fibers
      Also called "fight-or-flight"
      Cholinergic preganglionic fibers
      Adrenergic postganglionic fibers
      Dilates pupils of the eyes
      Increases heart rate
      Dilates blood vessels to skeletal muscles
   Checks (✓) for Parasympathetic
      Has terminal ganglia
      Also called the craniosacral division
      Cholinergic preganglionic fibers
      Also called "rest-and-repose" system
      Has short postganglionic fibers
      Cholinergic postganglionic fibers
      Increases digestive enzymes
      Constricts the bronchi

☞ **Answers to Chapter Self-Quiz**

1. d
2. Myelin
   Dendrite
   Microglia
   Oligodendrocyte
   Dendrite
3. 4 membrane becomes permeable to K
   1 threshold stimulus is applied
   6 sodium/potassium pump restores resting
   3 inside of membrane is positive to outside
   2 membrane becomes permeable to sodium
   5 membrane is polarized with potassium out

4. b
5. Dura mater
   Choroid plexus
   Third
   Endomysium
   Corpus callosum
6. b
7. B (arbor vitae)
   D (pons)
   A (insula)
   A (primary motor area)
   A (basal ganglia)
   C (thalamus)
   B (vermis)
   D (midbrain)
   A (lateral ventricles)
   D (cardiac center)
8. Trigeminal
   Optic
   Facial
   Vagus
   Hypoglossal
9. P (terminal ganglia)
   B (two neurons in pathway)
   B (cholinergic fibers)
   S (short preganglionic fibers)
   S (adrenergic fibers)
   P (conserves energy)
   P (craniosacral division)
   S (widespread and long lasting effect)
   P (shows little divergence)
   S (dilates blood vessels to skeletal muscles)

☞ **Answers to Terminology Exercises**

Glossopharyngeal
Anesthesia
Astrocyte
Encephalitis
Aphasia

Toward
Nerve glue
Knot or group of neuron cell bodies
Furrows between convolutions of cerebrum
Inflammation of meninges

E (paralysis of four extremities)
D (network of nerves)
B (tree-like processes)
C (membrane of a nerve fiber)
E (stiff)

☞ **Answers to Fun and Games**

Falx cerebri
Saltatory
Schwann
White
Dendrite
Cerebellum
White
Widespread
Neuroglia
Synapse
Pons
Vermis
Ganglion
Occipital
Pia mater

QUOTATION: All wish to possess knowledge but few, comparatively speaking, are willing to pay the price.

☞ **Quiz/Test Questions**

Note: There are fifty multiple-choice questions for this chapter in the computerized test bank.

Name the following:

1. Two components of the central nervous system.
   Answer: brain and spinal cord.

2. Afferent process of a neuron.
   Answer: dendrite.

3. Gap between myelin in a myelinated nerve fiber.
   Answer: node of Ranvier.

4. Neuroglia cell that is phagocytic.
   Answer: microglia.

5. Neuroglia cell that binds blood vessels to neurons and forms the blood brain barrier.
   Answer: astrocyte.

6. Predominant cation in the extracellular fluid outside a nonconducting neuron.
   Answer: sodium

7. Type of circuit in which a single presynaptic neuron synapses with many postsynaptic neurons.
   Answer: divergence.

8. Lobe of the cerebrum that contains the primary motor area.
   Answer: frontal lobe.

Questions for this chapter continue on page 138.

9 Sensory System

☞ **Key Terms/Concepts**

Accommodation Mechanism that allows the eye to focus at various distances, primarily achieved through changing the curvature of the lens.

Ampulla saclike dilation of a tube or duct; found at the end of each semicircular canal and contains the receptors for dynamic equilibrium.

Auditory ossicles Three tiny bones in the middle ear that function to amplify sound waves; malleus, incus, stapes.

Bulbus oculi The eyeball.

Chemoreceptor A sensory receptor that detects the presence of chemicals; responsible for taste, smell, and monitoring the concentration of certain chemicals in body fluids.

Cochlea The spiral, or coiled, portion of the bony labyrinth.

Dynamic equilibrium Equilibrium of motion; maintaining balance when the head or body is moving.

General senses Senses that are not localized, but are found throughout the body; includes touch, pressure, temperature, and pain.

Labyrinth Complex series of interconnecting cavities within the inner ear.

Lacrimal apparatus The structures that produce and convey tears.

Macula The structure in the utricle and saccule that detects a change in position of the head; functions in static equilibrium.

Mechanoreceptor A sensory receptor that responds to a bending or deformation of the cell; examples include receptors for touch, pressure, hearing, and equilibrium.

Nociceptor A sensory receptor that responds to tissue damage; pain receptor.

Organ of Corti The organ of hearing, consisting of supporting cells and hair cells that rest on the basilar membrane and project into the endolymph of the cochlear duct; also called spiral organ.

Photoreceptor A sensory receptor that detects light; located in the retina of the eye.

Proprioception Sense of position, or orientation, and movement.

Refraction The bending of light as it passes from medium to another.

Semicircular canals Three curved passageways in the bony labyrinth of the inner ear, filled with perilymph and contain the membranous semicircular ducts.

Sensory adaptation Phenomenon in which some receptors respond when a stimulus is first applied but decrease their response if the stimulus is maintained; receptor sensitivity decreases with prolonged stimulation.

Special senses Senses that involve localized sensory organs that contain the receptors; taste, smell, sight, hearing, equilibrium.

Static equilibrium Sensing and evaluating the position of the head relative to gravity.

Thermoreceptor A sensory receptor that detects changes in temperature.

Vestibule A small space at the entrance to a passage; in the inner ear, the vestibule is located adjacent to the oval window, between the cochlea and the semicircular canals.

☞ **Chapter Objectives**
Upon completion of this chapter the student should be able to:
1. Distinguish between general senses and special senses.
2. Classify sense receptors into five groups.
3. Explain what is meant by sensory adaptation.
4. Describe the sense receptors for touch, pressure, proprioception, temperature, and pain.
5. Describe the structure of a taste bud as it relates to the sense of taste.
6. Identify and locate the four different taste sensations and follow the impulse pathway from stimulus to the cerebral cortex.
7. Locate the receptors for the sense of small and follow the impulse pathway to the cerebral cortex.
8. Discuss the relationship between the sense of taste and sense of smell.
9. List the bones that form the orbit of the eye.
10. Describe the protective features of the eye.
11. Describe the structure of the bulbus oculi and the significance of each component.

12. Explain the concept of light refraction and name the four refractive surfaces and media in the eye.
13. Explain how the eyes accommodate for near vision.
14. Identify the photoreceptor cells in the retina and describe the mechanism by which nerve impulses are triggered in response to light.
15. Trace the pathway of a visual impulse from where it is initiated to where it is interpreted in the cerebral cortex.
16. Distinguish between the outer ear, middle ear, and inner ear and describe the structure of each region.
17. Distinguish between the bony labyrinth and membranous labyrinth of the inner ear.
18. Describe the contribution each region of the ear makes to the sense of hearing.
19. Summarize the sequence of events in the initiation of auditory impulses.
20. Explain how differences in pitch and loudness are perceived.
21. Trace the pathway of auditory impulses from where they are initiated to where they are interpreted in the cerebral cortex.
22. Distinguish between static equilibrium and dynamic equilibrium.
23. Identify and describe the structure of the components of the ear involved in static equilibrium.
24. Identify and describe the structure of the components of the ear involved in dynamic equilibrium.
25. Summarize the events in the initiation of impulses for static equilibrium and for dynamic equilibrium.
26. Identify the cranial nerve that transmits impulses for static and dynamic equilibrium to the cerebral cortex.

☞ **Chapter Outline/Summary**

Receptors and Sensations (Objectives 1 - 3)
- Receptors for the general senses are widely distributed in the body.
- Receptors for the special senses are localized.
- Receptors are classified as chemoreceptors, mechanoreceptors, nociceptors, thermoreceptors, or photoreceptors.
- Sensory adaptation occurs when a continued stimulus decreases the sensitivity of the receptors.

General Senses (Objective 4)
- The receptors for touch and pressure are mechanoreceptors. Receptors for touch and pressure include free nerve endings, Meissner's corpuscles, and Pacinian corpuscles.
- Proprioception is the sense of position or orientation. Golgi tendon organs and muscle spindles, receptors for proprioception, are mechanoreceptors.
- Temperature changes are detected by thermoreceptors. Thermoreceptors are free nerve endings. Some are sensitive to heat, others to cold. Temperature extremes also stimulate pain receptors.
- Receptors for pain are called nociceptors. These receptors are free nerve endings that are stimulated by tissue damage.

Gustatory Sense (Objectives 5 and 6)
- Receptors for taste are chemoreceptors.
- Taste buds, the organs for taste, are located on the walls of the papillae on the surface of the tongue.
- Taste buds contain the taste cells and supporting cells. Taste hairs on the taste cells function as the receptors.
- Sweet, salty, sour, and bitter are the four taste sensations. Sweet is located at the tip of the tongue, salty is on the anterior sides of the tongue, sour is on the posterior sides of the tongue, and bitter is at the back of the tongue.
- Impulses for taste are transmitted along the facial nerve or the glossopharyngeal nerve to the sensory cortex of the parietal lobe.

Olfactory Sense (Objectives 7 and 8)
- The sense of smell is called olfaction.
- The receptors for the sense of small are chemoreceptors and are located in the olfactory epithelium of the nasal cavity.
- The olfactory neurons enter the olfactory bulb. Impulses are transmitted along the olfactory tracts to the olfactory cortex in the temporal lobe.
- The senses of taste and smell are closely related and complement each other.

Visual Sense (Objectives 9 - 15)
Protective features and accessory structures of the eye (Objectives 9 and 10)
- The bony orbit is formed by the frontal, lacrimal, ethmoid, maxilla, zygomatic, sphenoid, and palatine bones.
- Eyebrows, eyelids, and eyelashes help to protect the eye from foreign particles and irritants.
- The lacrimal apparatus produces tears that moisten and cleanse the eye. Tears contain an enzyme that helps destroy bacteria.

Structure of the eyeball (bulbus oculi) (Objective 11)

- The sclera and cornea are parts of the outermost layer, or fibrous tunic, of the eye. They give shape to the eye and the cornea refracts light rays.
- The middle, or vascular, tunic includes the choroid, ciliary body, and iris. The choroid absorbs excess light rays; the ciliary body changes the shape of the lens; and the iris regulates the size of the pupil.
- The retina is the innermost layer, or nervous tunic, of the eye. It contains the receptor cells.
- The lens, suspensory ligaments, and ciliary body form a partition that divides the interior of the eye into two cavities. The anterior cavity is filled with aqueous humor and the posterior cavity is filled with vitreous humor. Both the aqueous and vitreous humors refract light rays.

Pathway of light and refractive media (Objectives 12 and 13)

- Refraction is the bending of light rays as they travel between substances of differing optical densities.
- The refractive media in the eye are the cornea, aqueous humor, lens, and vitreous humor.
- In the normal relaxed eye, the four refractive media sufficiently bend the light rays from objects at least 20 feet away to focus on the retina.
- When the eyes accommodate for close vision, the ciliary muscle contracts, the suspensory ligaments become less taut, and the lens becomes more convex so light rays have sufficient refraction to focus on the retina.

Photoreceptors (Objective 14)

- Rods are the photoreceptors for black and white vision and for vision in dim light. Cones are the receptors for color vision and visual acuity. The rods and cones are located in the retina.
- Rods contain rhodopsin, which breaks down into opsin and retinal when it is exposed to light. This reaction triggers a nerve impulse.
- Cones, concentrated in the fovea centralis, function in a manner similar to rods. Color is possible because one type of cones is sensitive to green light, one is sensitive to blue light, and the third responds to red light.

Visual pathway (Objective 15)

- Visual impulses triggered by the rods and cones travel on the optic nerve to the optic chiasma.

- From the optic chiasma, the impulses travel on the optic tracts to the thalamus.
- From the thalamus, the impulses travel on optic radiations to the visual cortex in the occipital lobe.

Auditory Sense (Objectives 16 - 21)

Structure of the ear (Objectives 16 - 18)

- Outer ear
  - The external ear, which includes the auricle and external auditory meatus, ends at the tympanic membrane.
  - The auricle collects the sound waves and directs them toward the external auditory meatus, which serves as a passageway to the tympanic membrane.
- Middle ear
  - The middle ear contains three tiny bones called auditory ossicles. These bones are the malleus, incus, and stapes .
  - The auditory ossicles transmit sound vibrations from the tympanic membrane to the oval window.
- Inner ear
  - The inner ear consists of a bony labyrinth that surrounds a membranous labyrinth. It includes the vestibule, semicircular canals, and cochlea. The cochlea is the part that functions in hearing.
  - The membranous labyrinth of the cochlea is the cochlear duct, which contains endolymph.
  - The organ of Corti, located on the basilar membrane in the cochlear duct, contains the receptors for sound.

Physiology of hearing (Objectives 19 - 21)

- Sound waves cause vibration of the tympanic membrane. Auditory ossicles transmit the vibrations through the middle ear to the oval window.
- Movement of the oval window passes the vibrations to the perilymph in the scala vestibuli and scala tympani in the inner ear.
- Vibrations in the perilymph create corresponding oscillations in the vestibular and basilar membranes of the cochlear duct.
- As the basilar membrane moves up and down, the hairs on the hair cells of the organ of Corti rub against the tectorial membrane. Mechanical deformation of the hairs triggers the nerve impulses.
- The interpretation of pitch is mediated by the portion of the basilar membrane that vibrates and loudness is interpreted by the number of hair cells that are stimulated.

- Cranial nerve VIII transmits auditory impulses to the medulla oblongata. From there, the impulses travel to the thalamus, and then to the auditory cortex of the temporal lobe.

Sense of Equilibrium (Objectives 22 - 26)

Static equilibrium

- Static equilibrium occurs when the head is motionless. It is involved in evaluating the position of the head relative to gravity.
- The organ of static equilibrium is the macula, located within the utricle and saccule, which are portions of the membranous labyrinth inside the vestibule.
- Hairs on the receptor cells bend and trigger an impulse, which is transmitted to the central nervous system on Cranial nerve VIII.

Dynamic equilibrium

- Dynamic equilibrium is the equilibrium of motion and occurs when the head is moving.
- The receptors for dynamic equilibrium are located in the crista ampullaris within the ampullae at the base of the semicircular canals.
- Hairs on the receptor cells bend and trigger an impulse, which is transmitted to the central nervous system on Cranial nerve VIII.

☞ Answers to Review Questions

1. Receptors for the general senses are widely distributed throughout the body. The receptors for the special senses are localized in a specific region.
2. Chemoreceptors, mechanoreceptors, nociceptors, thermoreceptors, and photoreceptors.
3. Sensory adaptation is a decreased sensitivity to a continued stimulus.
4. Meissner's corpuscles, free nerve endings, and pacinian (lamellated) corpuscles.
5. Proprioception.
   Golgi tendon organs and muscle spindles function in proprioception.
6. Tissue damage stimulates nociceptors.
7. Receptors for taste are in the taste buds located on the papillae on the surface of the tongue.
8. Salty: Margins of tip and sides of tongue
   Sweet: Tip of tongue
   Sour: Sides of tongue
   Bitter: Back of tongue
9. Cranial nerve I, Olfactory nerve
10. The receptors for both taste and smell are chemoreceptors. They complement each other and often have a combined effect when they are interpreted in the cerebral cortex.
11. Frontal, lacrimal, ethmoid, maxilla, zygomatic, sphenoid, and palatine.
12. Eyebrows help to keep perspiration, an irritant, out of the eyes.
    Eyelids open and close the eye and keep foreign particles out of the eye.
    Eyelashes help to trap foreign particles.
13. Tears are produced in the lacrimal gland, which is located along the upper lateral margin of the orbit. They moisten, cleanse, and lubricate the anterior surface of the eye and help to prevent infection by destroying bacteria through the action of lysozymes.
14. Should include:
    Fibrous tunic: sclera and cornea
    Vascular tunic: choroid, ciliary body, iris
    Nervous tunic: retina
15. Vitreous humor is located in the posterior cavity, behind the lens. Aqueous humor is located in the anterior cavity, anterior to the lens.
16. Cornea, aqueous humor, lens, vitreous humor.
17. When the ciliary muscle in the ciliary body contracts, tension on the suspensory ligaments is reduced, and the lens becomes more convex to provide refraction for the light rays.
18. Rods and cones are photoreceptor cells located in the retina. The rods are for black and white vision and for vision in dim light. Cones are for color vision and acuity.
19. The rods and cones contain light sensitive pigments. Rods contain rhodopsin, which is highly sensitive and decomposes in dim light. When rhodopsin breaks down into opsin and retinal, the reaction triggers a nerve impulse. Cones function in the same way, but there are three types of pigments, each sensitive to a particular wave length of light to give color vision. The three pigments are sensitive to red light, blue light, and green light.
20. Cranial nerve II, optic nerve, transmits visual impulses. The impulses are interpreted in the visual cortex of the occipital lobe.
21. See Figure 9-11 in textbook.
22. Malleus, incus, stapes.
23. Perilymph.
24. Auditory receptors are located in the organ of Corti within the cochlea of the inner ear.
25. Vibrations are passed from the tympanic membrane, through the bones of the middle ear, oval window, and finally to the perilymph of the inner ear. Oscillations of

the perilymph cause vibrations in the vestibular and basilar membranes. When the basilar membrane moves, the hairs on the hair cells in the organ of Corti rub against the tectorial membrane and bend. Bending of the hairs stimulates the formation of impulses.

26. Loudness is interpreted by the number of hair cells that are stimulated.

27. Auditory cortex is in the temporal lobe.

28. Static equilibrium involves evaluating the position of the head relative to gravity and occurs when the head is motionless or moving in a straight line. Dynamic equilibrium occurs when the head is moving in an angular or rotational direction.

29. The receptors for static equilibrium are located macula, which is in the membranous utricle and saccule within the vestibule of the inner ear.

30. The receptors for dynamic equilibrium are located in the crista ampullaris within the ampullae at the base of the semicircular canals.

31. A nerve impulse is triggered when the hairs on the hair cells within the receptors bend.

32. Vestibular branch of cranial nerve VIII, the vestibulocochlear nerve.

☞ **Answers to Learning Exercises**

Receptors and Sensations (Objectives 1-3)

1. Senses with receptors that are widely distributed within the body are called **general** senses. If the receptors are localized in a specific region, they are called **special**.

2. Photoreceptor
Baroreceptor
Thermoreceptor
Nociceptor
Chemoreceptor

3. Sensory adaptation occurs when certain receptors are continually stimulated and no longer respond unless the stimulus becomes more intense.

General Senses (Objective 4)

1. Check (✓) before each of the following:
Pacinian corpuscles - mechanoreceptors
Mechanoreceptors - proprioception
Pain - nociceptors

2. **Thermoreceptors** exhibit rapid sensory adaptation.
Nociceptors may send impulses after the stimulus is removed.

Gustatory Sense (Objectives 5 and 6)

1. Gustatory
Papillae
Taste hair cells
Chemoreceptors

2. Sweet, sour, salty, bitter
Facial
Glossopharyngeal
Parietal lobe

Olfactory Sense (Objectives 7 and 8)

1. Nasal cavity
2. I, temporal
3. taste, smell

Sense of Vision (Objectives 9-15)

1. Lacrimal
Conjunctiva
Sclera
Cornea
Choroid
Pupil
Aqueous humor
Iris
Ciliary
Produce aqueous humor
Retina
Rods and cones
Blind spot (optic disk)
Fovea centralis
Vitreous humor

2. Refraction

3. Accommodation is the adjustment needed to focus light rays for **close** vision. To focus light rays from close objects, the ciliary muscle **contracts**, which **decreases** the tension on the suspensory ligaments. When this happens, the lens becomes **thicker** and light rays are bent **more**. In addition, the pupil **constricts**.

4. C (anterior cavity
H (choroid)
F (ciliary body)
D (cornea)
A (eyelid)
B (iris)
K (lens)
L (optic nerve)
J (posterior cavity)
I (retina)
G (sclera)
E (suspensory ligaments)

5. Cones
Rhodopsin
Cones
Red, blue, green
Rods
Rhodopsin

Rods
Vitamin A
Optic disk
Rods
6. 3 (optic chiasma)
   4 (optic tract)
   6 (optic radiations)
   2 (optic nerve)
   5 (thalamus)
   7 (occipital lobe)
   1 (photoreceptors)

Sense of Hearing (Objectives 16-21)
1. F (ampulla)
   E (auditory tube)
   J (auricle)
   D (cochlea)
   I (external auditory canal)
   G (incus)
   K (malleus)
   A (semicircular canals)
   L (stapes)
   H (tympanic membrane)
   B (vestibule)
   C (vestibulocochlear nerve)
2. Bony labyrinth
   Membranous labyrinth
   Endolymph
   Perilymph
   Cochlea
   Cochlear duct
   Basilar membrane
   Vestibular membrane
   Basilar membrane
3. 1) Tympanic membrane
   2) Malleus
   3) Incus
   4) Stapes
   5) Perilymph
   6) Endolymph
   7) Hair cells
   8) Tectorial membrane
   9) Basilar membrane
   10) Oscillation

Sense of Equilibrium (Objectives 22-26)
1. Dynamic equilibrium
   Static equilibrium
   Macula
   Utricle
   Saccule
   Otoliths
   Crista ampullaris
   Ampulla of semicircular canals

☞ **Answers to Chapter Self-Quiz**

1. S P (vision)
   G M (touch)

S C (taste)
S C (smell)
G T (cold shower)
G N (toothache)
S M (static equilibrium)
G M (pressure)
G N (pain from a burn)
S M (hearing)
2. c
3. b
4. d
5. Retina
   Sclera
   Aqueous humor
   Ciliary muscle
   Fovea centralis
   Cones
   Iris
   Optic disk
   Cornea
   Rhodopsin
6. H (malleus, incus, stapes)
   B (membranous labyrinth)
   E (semicircular canals)
   H (cochlea)
   E (utricle and saccule)
   H (tectorial membrane)
   B (endolymph)
   B (hair cells)
   H (organ of Corti)
   B (vestibulocochlear nerve)
7. Organ of Corti
   Macula
   Cristae ampularis

☞ **Answers to Terminology Exercises**

Audiology
Otolith
Vitreous humor
Gustatory
Cochlea

Pertaining to smell
Yellow spot
Inside the eye
Surgical removal of the eardrum
Pertaining to tears

E (hearing of old age)
C (inflammation of the cornea)
D (surgical repair of the ear)
B (surgical excision of a portion of the iris)
A (central pit)

☞ Answers to Fun and Games

QUOTATION: An eye can threaten like a loaded and levelled gun, or it can insult like hissing or kicking, or in its altered mood, by beams of kindness, it can make the heart dance for joy.

☞ Quiz/Test Questions

Note: There are fifty multiple-choice questions for this chapter in the computerized test bank.

Name the following:

1. Receptors that detect tissue damage.
   Answer: nociceptors.

2. Two special senses that depend on chemoreceptors.
   Answer: taste and smell.

3. Gland that produces tears in the eyes.
   Answer: lacrimal gland.

4. Outermost tunic of the eye in the posterior region.
   Answer: sclera.

5. Area of the retina where vision is sharpest.
   Answer: fovea centralis.

6. Pigmented portion of the eye that has radial and circular muscles.
   Answer: iris.

7. Fluid in the anterior cavity of the eye.
   Answer: aqueous humor.

8. Refractive component of the eye that adjusts for close vision.
   Answer: lens.

9. Region where some of the optic nerve fibers cross over to the opposite side.
   Answer: optic chiasma.

10. Three bones in the middle ear cavity.
    Answer: malleus, incus, stapes.

11. Region of the inner ear that contains the organ of hearing.
    Answer: cochlea.

12. Type of receptors involved in hearing and equilibrium.
    Answer: mechanoreceptors.

13. Structure that rests on the basilar membrane and contains the hair cells for hearing.
    Answer: organ of Corti.

14. Portions of the membranous labyrinth that contain the receptors for static equilibrium.
    Answer: utricle and saccule.

15. Organ that contains the hair cells for the sense of dynamic equilibrium.
    Answer: cristae ampullaris.

True/False Questions:

1. Proprioception is the sense of position.
   Answer: True.

2. Contraction of ciliary muscles causes the lens to flatten for distance vision.
   Answer: False; contraction of the ciliary muscles causes the lens to become more convex for distance vision.

3. Rods are most numerous in the fovea centralis where vision is sharpest.
   Answer: False; cones, for sharp vision, are most numerous in the fovea centralis.

4. Visual impulses travel along the optic nerve to the optic chiasma, then along optic tracts to the thalamus, and then along optic radiations to the frontal lobe of the cerebral cortex where they are interpreted.
   Answer: False; visual impulses are interpreted in the occipital lobe.

5. The auditory ossicle attached to the tympanic membrane is the malleus.
   Answer: True.

6. Fluid within the bony labyrinth, but outside the membranous labyrinth, of the inner ear is called perilymph.
   Answer: True.

7. The region of the basilar membrane near the base of the cochlea vibrates in response to high frequency sound waves.
   Answer: True.

8. Otoliths are calcium carbonate crystals embedded in the gelatinous mass of the cupula in the crista ampullaris.
   Answer: False; they are embedded in the gelatinous mass on the macula within the utricle and saccule.

9. The light sensitive pigment in the rods is rhodopsin.
   Answer: True.

10. The crista ampullaris is involved in detecting the position of the head relative to gravity or movement in a straight line.
    Answer: False; the crista ampullaris is involved in dynamic equilibrium or rotational movement.

10 Endocrine System

☞ **Key Terms/Concepts**

Adenohypophysis Anterior portion of the pituitary gland.

Adrenal cortex Outer portion of the adrenal gland that secretes hormones called corticoids.

Adrenal medulla Inner portion of the adrenal gland that secretes epinephrine and norepinephrine.

Endocrine gland A gland that secretes its product directly into the blood; a ductless gland.

Exocrine gland A gland that secretes its product to a surface or cavity through ducts.

Gonads Sex organs in which the reproductive cells are formed; testes in men and ovaries in women.

Hormone A substance secreted by an endocrine gland.

Negative feedback A mechanism of response in which a stimulus initiates reactions that reduce the stimulus.

Neurohypophysis Posterior portion of the pituitary gland.

Pancreas A glandular organ in the abdominal cavity that has both exocrine and endocrine functions; the exocrine portion consists of acinar cells, the endocrine portion is the islets of Langerhans.

Pancreatic islets Endocrine portion of the pancreas; consist of alpha cells that secrete glucagon and beta cells that secrete insulin.

Parathyroid glands A set of small glands embedded on the posterior aspect of the thyroid gland.

Pineal gland A region of the epithalamus in the diencephalon that is thought to be involved with regulating the "biological clock"; also called the pineal gland because it secretes melatonin.

Pituitary gland Endocrine gland located in the sella turcica of the sphenoid bone, near the base of the brain; also called the hypophysis.

Prostaglandins A group of substances, derived from fatty acids, that are produced in small amounts and have an immediate, short term, localized effect; sometimes called local hormones.

Suprarenal gland Endocrine gland that is located on the superior pole of each kidney; divided into cortex and medulla regions; also called the adrenal gland.

Target tissue A tissue (cells) that responds to a particular hormone because it has receptor sites for that hormone.

Thymus Endocrine gland located in the mediastinum; secretes thymosin; plays an important role in the body's immune system.

Thyroid gland Endocrine gland that is located anterior to the trachea at the base of the neck.

☞ **Chapter Objectives**

Upon completion of this chapter the student should be able to:

1. Compare the actions of the nervous system and the endocrine system.
2. Distinguish between the characteristics of exocrine glands and endocrine glands.
3. Identify two chemical classes of hormones.
4. Explain why hormones affect only target tissues and not other tissues in the body.
5. Distinguish between receptor site locations for protein and steroid hormones.
6. Identify three different mechanisms for regulating hormone secretion.
7. Describe the mechanism of negative feedback regulation.
8. Identify the eight major endocrine glands.
9. Compare the anterior and posterior lobes of the pituitary gland in terms of embryonic origin, mechanisms that regulate their activity, and the hormones that are secreted.
10. Name six hormones secreted by the adenohypophysis and describe the action of each hormone.
11. Name two hormones secreted by the neurohypophysis and describe the action of each hormone.
12. Describe the location, structure, and hormones of the thyroid gland.
13. Discuss the location and actions of the parathyroid gland.
14. Compare the actions of calcitonin and parathyroid hormone.
15. Describe the location, structure, and regulation of the adrenal glands.

16. Identify the three groups of hormones that are secreted by the cortex of the adrenal gland and describe the functions of each group.

17. Discuss the physiologic effects of hypersecretion and hyposecretion of the adrenocortical hormones.

18. Name two hormones from the adrenal medulla and describe their effects.

19. Describe the location and structure of the pancreas.

20. Compare the source and actions of glucagon and insulin.

21. Identify the principle androgen and state its general function.

22. Name two hormones produced by the ovaries and state their general functions.

23. Describe the location of the pineal gland and discuss its endocrine function.

24. Discuss the action of the thymus gland.

25. Name and describe the function of at least one hormone from the (a) gastric mucosa; (b) small intestine; (c) heart; and (d) placenta.

26. Differentiate between hormones and prostaglandins.

☞ Chapter Outline/Summary

Introduction to the Endocrine System (Objectives 1 and 2)

Comparison of endocrine system and nervous system (Objective 1)

- The nervous system acts through electrical impulses and neurotransmitters; the effect is localized and of short duration.
- The endocrine system acts through chemicals called hormones; the effect is generalized and of long term duration.

Comparison of endocrine glands and exocrine glands (Objective 2)

- Exocrine glands have ducts that carry their product to some surface.
- Endocrine glands are ductless, their products, called hormones, are secreted directly into the blood.

Characteristics of Hormones (Objectives 3 - 7)

Chemical nature of hormones (Objective 3)

- Hormones are classified chemically as either proteins or steroids.
- Most hormones in the body are proteins or protein derivatives.
- The sex hormones and those from the adrenal cortex are steroids.

Mechanism of hormone action (Objectives 4 and 5)

- Hormones react with receptor sites on selected cells.
- The cells that have receptor sites for a specific hormone make up the target tissue for that hormone.
- Protein hormones react with receptors on the surface of the cell; steroids react with receptors inside the cell.

Control of hormone action (Objectives 6 and 7)

- Many hormones are regulated by a negative feedback mechanism.
- Some hormones are secreted in response to other hormones.
- A third method for regulating hormone secretion is by direct nerve stimulation.

Endocrine Glands and their Hormones (Objectives 8-25)

Pituitary gland (Objectives 9-11)

- The pituitary gland is divided into an anterior lobe, or adenohypophysis, which is regulated by releasing hormones from the hypothalamus, and a posterior lobe, or neurohypophysis, which is regulated by nerve stimulation.
- Hormones of the anterior lobe (adenohypophysis) (Objective 10)
    ○ Growth hormone (GH), or somatotropic hormone (STH) promotes protein synthesis, which results in growth.
    ○ Thyroid stimulating hormone (TSH) stimulates the activity of the follicular cells of the thyroid gland.
    ○ Adrenocorticotropic hormone (ACTH) stimulates activity of the adrenal cortex, particularly the secretion of cortisol.
    ○ Follicle stimulating hormone (FSH) is a gonadotropin that stimulates the development of ova in the ovaries and sperm in the testes. It also stimulates the production of estrogens in the female.
    ○ Luteinizing hormone (LH), another gonadotropin, causes ovulation and the secretion of progesterone and estrogens in the female. In the male it stimulates the production of testosterone.
    ○ Prolactin promotes the development of glandular tissue in the breast and stimulates the production of milk.
- Hormones of the posterior lobe (neurohypophysis) (Objective 11)
    ○ Antidiuretic hormone (ADH) promotes reabsorption of water in the kidney tubules.
    ○ Oxytocin causes uterine muscle contraction and ejection of milk from the lactating breast.

Thyroid gland (Objective 12)

- Thyroxine and triiodothyronine
  - Both thyroxine and triiodothyronine require iodine for synthesis.
  - About 95% of active thyroid hormone is thyroxine.
  - Thyroid hormone secretion is regulated by a negative feedback mechanism that involves the amount of circulating hormone, the hypothalamus, and TSH from the adenohypophysis.
  - Thyroid hormones affect the metabolism of carbohydrates, proteins, and lipids. Hyperthyroidism is characterized by a high metabolic rate. Hypothyroidism leads to conditions related to decreased metabolism.
- Calcitonin
  - Calcitonin is produced by parafollicular cells in the thyroid gland.
  - Calcitonin reduces calcium levels in the blood.

Parathyroid glands (Objective 13 and 14)

- Parathyroid glands are embedded on the posterior surface of the thyroid gland.
- Parathyroid hormone, antagonistic to calcitonin, increases blood calcium levels.
- Hypoparathyroidism reduces blood calcium levels, which may result in nerve irritability. Hyperparathyroidism results in calcium loss from bones and precipitation in abnormal places.

Adrenal (Suprarenal) gland (Objectives 15-18)

- Hormones of the cortex (Objectives 16 and 17)
  - All hormones from the adrenal cortex are steroids.
  - The adrenal cortex is regulated by a negative feedback mechanism involving the hypothalamus and ACTH from the adenohypophysis.
  - Hormones from the adrenal cortex are classified as either mineralocorticoids, glucocorticoids, or sex steroids.
  - The principal mineralocorticoid is aldosterone, which promotes sodium ion reabsorption in the kidney tubules.
  - The principal glucocorticoid is cortisol, which increases blood glucose levels and helps to counteract the inflammatory response.
  - Sex steroids from the adrenal cortex have minimal effect compared to the hormones from the ovaries and testes.

- Hormones of the medulla (Objective 18)
  - The two hormones produced by the adrenal medulla are epinephrine and norepinephrine. About 80% of the product is epinephrine.
  - Epinephrine and norepinephrine prepare the body for strenuous activity and stress.
  - The effect of epinephrine and norepinephrine is similar to that of the sympathetic nervous system but lasts up to ten times longer.

Pancreas -- Islets of Langerhans (Objectives 19 and 20)

- The endocrine portion of the pancreas consists of the pancreatic islets, or Islets of Langerhans.
- Alpha cells in the islets produce glucagon, which raises blood glucose levels.
- Beta cells produce insulin, which is antagonistic to glucagon, and decreases blood glucose levels.

Gonads (Objectives 21 and 22)

- Testes (Objective 21)
  - The testes produce the male sex hormones, which are collectively called androgens.
  - The principal androgen is testosterone, which is responsible for the development and maintenance of male secondary sex characteristics.
- Ovaries (Objective 22)
  - The ovaries produce estrogens and progesterone.
  - Estrogens are responsible for the development and maintenance of female secondary sex characteristics.
  - Progesterone maintains the uterine lining for pregnancy.

Pineal Gland (Body) (Objective 23)

- The pineal gland extends posteriorly from the third ventricle of the brain. Its secretory cells, called pinealocytes, synthesize and secrete the hormone melatonin.
- Melatonin inhibits gonadotropin releasing hormone from the hypothalamus, which inhibits reproductive functions.
- Melatonin also regulates circadian rhythms.

Other Endocrine Glands (Objectives 24 and 25)

- The thymus, located near the midline in the anterior portion of the thoracic cavity, produces the hormone thymosin. Thymosin plays an important role in the development of the body's immune system.

- The gastric mucosa secretes gastrin, which stimulates the secretion of hydrochloric acid and pepsin for the digestion of food.
- The mucosa of the small intestine secretes cholecystokinin and secretin. Cholecystokinin stimulates the pancreas to secrete digestive enzymes and contraction of the gallbladder. Secretin promotes the production of a bicarbonate-rich fluid in the pancreas.
- The heart produces a hormone called atrial natriuretic hormone, or atriopeptin, which promotes the loss of sodium and water in the urine to decrease blood pressure.
- The placenta is a temporary endocrine gland that secretes human chorionic gonadotropin which helps to maintain the uterine lining during pregnancy. It also produces estrogen and progesterone.

Prostaglandins (Objective 26)
- Prostaglandins are hormone-like molecules that are derived from arachidonic acid.
- They are produced by cells widely distributed throughout the body, their effect is localized near their origin, and their effect is immediate and short term.
- Prostaglandins are not stored; they are synthesized when needed.

☞ **Answers to Review Questions**

1. The nervous system acts through electrical impulses and neurotransmitters; the effect is localized and of short duration. The endocrine system acts through chemicals called hormones and is effect is more generalized and long term.
2. Exocrine glands have ducts that carry their product to a surface. Endocrines secrete their products directly into the blood, which transports it to the target tissue.
3. Proteins and steroids.
4. The target tissue cells have receptor sites for the hormone.
5. Receptor sites for protein hormones are on the cell membrane. Receptor sites for steroid hormones are inside the cell.
6. Negative feedback, other hormones, and nervous stimulation.
7. In negative feedback regulation, a gland is sensitive to the concentration of a substance it regulates. If a hormone from a gland decreases the concentration of a substance in the blood, then when that substance decreases, the gland is inhibited or the stimulus to produce the hormone is removed.

8. Pituitary gland, thyroid gland, parathyroid gland, adrenal gland, pancreas, pineal gland, thymus, and gonads
9. The anterior lobe of the pituitary gland, or adenohypophysis, develops from the embryonic oral cavity. The posterior lobe, or neurohypophysis, develops as an extension of the brain.
10. The activity of the adenohypophysis is controlled by releasing hormones from the hypothalamus. The neurohypophysis is controlled by neural stimulation.
11. Growth hormone, thyroid-stimulating hormone, adrenocorticotropic hormone, follicle-stimulating hormone, luteinizing hormone, and prolactin.
12. **Growth hormone** stimulates growth by promoting protein synthesis.
 Thyroid-stimulating hormone increases the secretion of thyroid hormone.
 Adrenocorticotropic hormone increases the secretion of hormones from the adrenal cortex, especially cortisol.
 Follicle-stimulating hormone promotes follicle maturation and estrogen secretion in the female and in the male it stimulates spermatogenesis.
 Luteinizing hormone stimulates ovulation and progesterone production in the female and in the male it stimulates the production of testosterone.
 Prolactin stimulates milk production.
13. Antidiuretic hormone and oxytocin are produced by specialized cells in the hypothalamus and are secreted from the neurohypophysis. Antidiuretic hormone increases water reabsorption in the kidney tubules to decrease the amount of urine. Oxytocin stimulates uterine contractions and the ejection of milk from the mammary glands.
14. The thyroid follicles secrete iodine-containing hormones.
15. A deficiency of iodine causes a simple goiter. Without iodine the thyroid cannot synthesize active hormone so the anterior pituitary secretes more TSH. The thyroid enlarges in an attempt to make more hormone, but the iodine isn't available.
16. Hypersecretion of thyroxine results in a high metabolic rate. Hyposecretion results in a low metabolic rate, lethargy, weight gain, slow heart rate. In the untreated infant, hyposecretion results in abnormally proportioned mentally retarded dwarf, called a cretin.

17. Calcitonin is produced by the parafollicular cells of the thyroid. It decreases blood calcium levels.

18. The parathyroid glands are embedded on the posterior surface of the thyroid gland. Parathyroid hormone is antagonistic to calcitonin and increases blood calcium levels.

19. Calcitonin and parathyroid hormone are antagonistic to each other.

20. The adrenal medulla is controlled by direct nerve stimulation. The adrenal cortex is regulated by ACTH from the anterior pituitary, which is secreted in response to releasing hormones from the hypothalamus.

21. **Mineralocorticoids**, primarily aldosterone, regulate sodium balance, by increasing the reabsorption of sodium in the kidneys. This usually causes water retention and secretion of potassium.
    Glucocorticoids, primarily cortisol, increase blood glucose levels and help to counteract the inflammatory response.
    Gonadocorticoids, primarily androgens, are sex steroids, but have minimal effect compared with the hormones from the gonads.

22. The mineralocorticoids are essential to life because they regulate fluid and electrolyte balance and sodium balance is critical.

23. The hormones of the adrenal medulla, epinephrine and norepinephrine, have an effect similar to the sympathetic nervous system.

24. Pancreatic islets or islets of Langerhans.

25. Alpha cells produce glucagon and beta cells produce insulin.

26. Glucagon increases blood glucose levels and insulin decreases blood glucose levels.

27. The main source of testosterone is the interstitial cells of the testes.

28. Estrogen and progesterone.

29. Melatonin has a regulatory role in sexual and reproductive development. It also functions in the organization and regulation of daily physiologic cycles.

30. The thymus has an important role in the development of the body's immune system.

31. **Gastrin**: gastric (stomach) mucosa
    Secretin: mucosa of small intestine
    Cholecystokinin: mucosa of small intestine
    Atrial natriuretic hormone: heart
    Human chorionic gonadotropin: placenta

32. Prostaglandins are produced in minute amounts by widely distributed cells. They have an immediate, short-term, localized effect and cannot be stored in the body.

33. Some effects of prostaglandins are to modulate hormone action, affect smooth muscle contraction, affect blood clotting, promote the inflammatory response.

☞ Answers to Learning Exercises

Introduction to the Endocrine System (Objectives 1 and 2)

1. E (acts through hormones)
   N (effect is localized)
   N (acts through electrical impulses)
   N (effect is of short term duration)
   E (effect is generalized and long term)

2. Exocrine glands have **ducts** that carry the secretory product to a surface. In contrast, **endocrine glands** are ductless and secrete their product directly into the **blood** for transport to the target tissue.

Characteristics of Hormones (Objective 3-7)

1. Protein
   Steroids
   Target cells
   Cell membrane
   Cytoplasm
   Neural
   Humoral (negative feedback)
   Hormonal

2. 1) Protein
   2) Adenyl cyclase
   3) Cyclic AMP
   4) ATP
   5) Hormone
   6) Cyclic AMP
   7) Steroid
   8) Receptor
   9) Nucleus
   10) DNA

Endocrine Glands and their Hormones (Objective 8-25)

1. D (adrenal gland)
   E (ovaries)
   H (pancreas)
   B (parathyroid)
   A (pineal gland)
   F (pituitary gland)
   C (thymus gland)
   G (thyroid gland)
   I (testes)

2. B (secretes ADH)
   A (stimulated by releasing hormones)
   A (secretes prolactin)
   B (derived from nervous tissue)
   B (secretes oxytocin)
   A (derived from embryonic oral cavity)

A (secretes TSH)
A (secretes growth hormone)
A (secretes ACTH)
B (regulated by nerve stimulation)
3. Thyroid-stimulating hormone
   Adrenocorticotropic hormone
   Luteinizing hormone
   Oxytocin
   Antidiuretic hormone
   Follicle-stimulating hormone
   Growth hormone (somatotropin)
   Oxytocin
   Luteinizing hormone (ICSH)
   Prolactin
4. C (secreted by parafollicular cells)
   T (requires iodine for production)
   P (increases blood calcium levels)
   T (secreted by thyroid follicles)

   T (increases rate of metabolism)
   C (reduces blood calcium levels)
   P (increases osteoclast activity)
   P (hyposecretion leads to nerve excitability)
5. Medulla
   Mineralocorticoids
   Aldosterone
   Cortex
   Cortisol
   Aldosterone
   Cortisol
   Gonadocorticoids
   Epinephrine and norepinephrine
6. Insulin
   Glucagon
   Glucagon
   Insulin
   Insulin
   Glucagon
   Glucagon
   Insulin
   Insulin
7. G 7 (testosterone)
   C 10 (melatonin)
   H 4 (thymosin)
   B 5 (estrogen)
   D 6 (HCG)
   F 9 (gastrin)
   E 1 (secretin)
   B 8 (progesterone)
   E 2 (cholecystokinin)
   A 3 (atrial peptin)

Prostaglandins (Objective 26)
1. Prostaglandins are similar to hormones but they are different in many ways. They are derivatives of **arachidonic acid** and the cells that produce them are **scattered (distributed)** throughout the body.
2. In contrast to hormones, the effects of prostaglandins are **localized**, **immediate**, and **short term**.

☞ **Answers to Chapter Self-Quiz**

1. P (growth hormone)
   P (epinephrine)
   S (aldosterone)
   P (insulin)
   S (cortisol)
   P (follicle-stimulating hormone)
2. A target tissue includes all the cells and/or tissues that have receptor sites for a given hormone.
3. A (growth hormone)
   A (gonadotropins)
   C (epinephrine)
   I (melatonin)
   A (luteinizing hormone)
   D (oxytocin)
   B (aldosterone)
   A (prolactin)
   E (progesterone)
   G (insulin)
   L (calcitonin
   D (antidiuretic hormone)
   B (cortisol)
   F (glucagon)
   J (testosterone)
   L (thyroxine)
   A (TSH)
   K (thymosin)
   E (estrogens)
   A (ACTH)
4. Follicle-stimulating hormone
   Prolactin
   Thymosin
   Thyroxine
   Luteinizing hormone
   Parathormone
   Melatonin
   Oxytocin
   Follicle-stimulating hormone
   Insulin
   Aldosterone
   Epinephrine
   Cortisol
   Luteinizing hormone (or ICSH)
   Calcitonin
5. P (derived from arachidonic acid)
   H (transported in the blood)
   H (may be stored in the body)

P (produced by cells...)
P (have a localized effect)
H (derived from proteins and lipids)

☞ **Answers to Terminology Exercises**

Oxytocin
Pineal
Adrenal
Progesterone
Diuretic

Tumor of a gland
Study of endocrine glands
Steroid from the testes
Beside or next to the thyroid
Influences the cortex of adrenal gland

C (assembles glucose into the blood)
A (surgical excision of the adrenal gland)
D (excessive thirst)
B (hormones that produce males or maleness)
E (hormone necessary before milk production)

☞ **Answers to Fun and Games**

1. Parathormone
2. Prolactin
3. Oxytocin
4. Calcitonin
5. Aldosterone
6. Testosterone
7. Insulin
8. Glucagon
9. Estrogen
10. Progesterone

☞ **Quiz/Test Questions**

Note: There are fifty multiple-choice questions for this chapter in the computerized test bank.

Name the following:

1. Secretory products of endocrine glands.
   Answer: hormones.

2. Chemical class of hormones that diffuse through the cell membrane and react with a receptor in the cytoplasm.
   Answer: steroids.

3. Second messenger in protein hormone action, the one that causes the effect attributed to the hormone.
   Answer: cyclic AMP.

4. Two gonadotropins from the adenohypophysis.
   Answer: follicle-stimulating hormone and luteinizing hormone.

5. Hormone from the neurohypophysis that affects kidney function.
   Answer: antidiuretic hormone (ADH).

6. Hormone that is antagonistic to calcitonin.
   Answer: parathyroid hormone or parathormone.

7. Portion of the adrenal gland that is under direct neural stimulation.
   Answer: adrenal medulla.

8. Hormone from the adrenal cortex that has an antiinflammatory effect.
   Answer: cortisol (glucocorticoid).

9. Two hormones that are antagonistic to each other and secreted by different cells of the same gland.
   Answer: insulin and glucagon.

10. Hormone that is stimulated by luteinizing hormone in the male.
    Answer: testosterone.

11. Ovarian hormone that is responsible for the development and maintenance of female secondary sex characteristics.
    Answer: estrogen.

12. Ovarian hormone that is stimulated by luteinizing hormone.
    Answer: progesterone.

13. Two hormones that on the mammary glands to cause milk production and ejection.
    Answer: prolactin and oxytocin.

14. Source of atrial natriuretic hormone or atriopeptin.
    Answer: heart.

15. Derivatives of arachidonic acid that are similar to hormones in many ways.
    Answer: prostaglandins.

True/False Questions:

1. Hormones from the neurohypophysis are regulated by releasing hormones from the hypothalamus.
   Answer: False; the adenohypophysis is regulated by releasing hormones from the hypothalamus.

2. Hyposecretion of growth hormone in the child usually leads to an abnormally proportioned and mentally retarded dwarf.
   Answer: False; the proportions and mental faculties are usually normal.

3. Follicle-stimulating hormone and luteinizing hormone are called gonadotropins.
   Answer: True.

Questions for this chapter continue on page 139.

11 Blood

Key Terms/Concepts

Agglutinin A specific substance in plasma that is capable of causing a clumping of red blood cells; an antibody.

Agglutinogen An genetically determined antigen on the cell membrane of a red blood cell that determines blood types.

Albumin The most abundant plasma protein, which is primarily responsible for regulating the osmotic pressure of the blood.

Basophil A white blood cell with granules in the cytoplasm that stain readily with basic dyes.

Coagulation The process of blood clotting.

Diapedesis The process by which white blood cells squeeze between the cells in a vessel wall to enter the tissue spaces outside the blood vessel.

Eosinophil A white blood cell with granules in the cytoplasm that stain readily with acid (eosin) dyes.

Erythrocyte Red blood cell.

Erythropoietin A hormone released by the kidneys that stimulates red blood cell production.

Fibrinogen A soluble plasma protein that is converted to insoluble fibrin by the action of thrombin during the process of blood clotting.

Formed elements Red blood cells, white blood cells, and platelets in the blood.

Globulin One type of proteins in the blood plasma.

Hemocytoblast A stem cell in the bone marrow from which the blood cells arise.

Hemopoiesis The process of blood cell formation in the red marrow of bones.

Hemostasis The stoppage of bleeding.

Leukocyte White blood cell.

Lymphocyte A type of white blood cell that lacks granules in the cytoplasm and has an important role in immunity.

Monocyte A type of white blood cell that lacks granules in the cytoplasm and is capable of phagocytosis.

Neutrophil A type of white blood cell that has granules in the cytoplasm that stain with acid and basic dyes and is capable of phagocytosis.

Plasma The liquid portion of blood.

Prothrombin A protein that is produced by the liver and released into the blood where it is converted to thrombin during the process of blood clotting.

Thrombocyte One of the formed elements of the blood, which functions in blood clotting; also called platelet.

Chapter Objectives

Upon completion of this chapter the student should be able to:

1. Describe five physical characteristics of blood.
2. List six functions of blood.
3. Identify the two parts of a blood sample and state the normal percentage of total volume for each one.
4. Describe the composition of blood plasma.
5. List three categories of formed elements in the blood.
6. Identify seven formed elements of the blood and state at least one function for each formed element.
7. Discuss the life cycle of erythrocytes.
8. Differentiate between five types of leukocytes and tell whether each one is an agranulocyte or a granulocyte.
9. Describe the three processes that constitute hemostasis.
10. Summarize the series of chemical reactions involved in the formation of a blood clot into three main steps.
11. Define fibrinolysis.
12. Define agglutinogen and agglutinin.
13. State the agglutinogens and agglutinins present in each of the four ABO blood types and explain why different blood types are incompatible for transfusions.
14. Explain the difference between Rh+ and Rh- blood.
15. Discuss the pathogenesis and treatment of hemolytic disease of the newborn.

Chapter Outline/Summary

Functions and characteristics of the blood
(Objectives 1 and 2)
- Blood is a liquid connective tissue; measures about 5 liters; accounts for 8% body weight;

slightly heavier than water; 4 to 5 times more viscous than water; pH is 7.35 to 7.45.

- Blood transports gases, nutrients, and waste products; helps regulate body temperature, fluid and electrolyte balance, pH; helps prevent fluid loss and disease.

Composition of the blood (Objectives 3 - 8)

- Blood is 55% plasma and 45% formed elements. (Objective 3)

Plasma (Objective 4)

- Plasma proteins
  - ○ Albumins account for 60% of the plasma proteins. They maintain the osmotic pressure of the blood.
  - ○ Globulins account for 36% of the plasma proteins. They function in lipid transport and in immune reactions.
  - ○ Fibrinogen accounts for 4% of the plasma proteins. It functions in the formation of blood clots.
- Nonprotein molecules that contain nitrogen, such as amino acids, urea, and uric acid may be present in blood plasma.
- Nutrients and gases
  - ○ Simple nutrients that are the end products of digestion are transported in the plasma.
  - ○ Oxygen and carbon dioxide are gases that are transported in the plasma.
- Electrolytes
  - ○ Sodium, potassium, calcium, chloride, bicarbonate, and phosphate ions are common electrolytes in the blood plasma.
  - ○ Electrolytes contribute to the osmotic pressure of the blood, maintain membrane potentials, and regulate pH of body fluids.

Formed elements (Objectives 5-8)

- Erythrocytes (Objectives 5-7)
  - ○ Erythrocytes are anucleate, biconcave disks, about 7.5 micrometers in diameter; there are 4.5 - 6 million/mm^3 blood; they contain hemoglobin. The primary function of erythrocytes is to transport oxygen.
  - ○ Erythrocyte production is regulated by erythropoietin, which is activated by renal erythropoietin factor. Iron, vitamin B$_{12}$, and folic acid are essential for RBC production.
  - ○ The life span of RBC's is about 120 days then they are destroyed by the spleen and liver. The iron and protein portions are reused; the pigment portion is converted to bilirubin and is secreted in bile.

- Leukocytes (Objectives 5, 6, and 8)
  - ○ Leukocytes have a nucleus and do not have hemoglobin; they average between 5,000/mm^3 and 9,000/mm^3; they move through capillary walls by diapedesis. WBC's provide a defense against disease and mediate inflammatory reactions.
  - ○ Neutrophils are granulocytes with light colored granules; they are the most numerous leukocyte and are phagocytic.
  - ○ Eosinophils are granulocytes with red granules; they help counteract the effects of histamine.
  - ○ Basophils are granulocytes with blue granules; they secrete histamine and heparin. In the tissues they are called mast cells.
  - ○ Lymphocytes are agranulocytes that have a special role in immune processes; some attack bacteria directly, others produce antibodies.
  - ○ Monocytes are large phagocytic agranulocytes. In the tissues they are called macrophages.
- Thrombocytes (Objective 5 and 6)
  - ○ Thrombocytes, or platelets, are fragments of megakaryocytes that function in blood clotting.
  - ○ Thrombocytes average 250,000 - 500,000/mm^3 of blood.

Hemostasis (Objectives 9-11)

- Hemostasis, the stoppage of bleeding, includes vascular constriction, platelet plug formation, and coagulation.
- Vascular constriction: The initial reaction in hemostasis is vascular constriction, which reduces the flow of blood through a torn or severed vessel.
- Platelet plug formation: Collagen from damaged tissues attracts platelets, which form a plug to fill the gap in a broken vessel to reduce blood loss.
- Coagulation (Objectives 10 and 11)
  - ○ Coagulation, or blood clot formation, starts with the formation of prothrombin activator, continues with the conversion of prothrombin to thrombin, and ends with the conversion of soluble fibrinogen to insoluble fibrin.
  - ○ Calcium and vitamin K are necessary for successful clot formation.
  - ○ After a clot forms, it condenses, or retracts, to pull edges of wound together.
  - ○ As healing takes place, the clot dissolves by fibrinolysis.

Blood Typing and Transfusions (Objectives 12-15)

Agglutinogens and agglutinins (Objective 12)
- Blood type antigens on the surface of RBC's are called agglutinogens.
- Antibodies that react with agglutinogens are in the plasma and are called agglutinins.

ABO blood groups (Objective 13)
- The ABO blood types are based on the agglutinogens present on the surface of the RBC's.
- Type A blood has type A agglutinogens and anti-B agglutinins; type B blood has type B agglutinogens and anti-A agglutinins; type AB blood has both type A and type B agglutinogens but neither agglutinin; type O blood has neither agglutinogen but has both anti-A and anti-B agglutinins.
- In transfusion reactions involving mismatched blood, the recipient's agglutinins react with the donor's agglutinogens.
- Type AB blood is called the universal recipient and type O is the universal donor.

Rh factor (Objectives 14 and 15)
- People that have Rh+ blood have Rh agglutinogens; Rh- individuals do not have Rh agglutinogens. Normally, neither type has anti-Rh agglutinins.
- Exposure to Rh+ blood causes an Rh- individual to develop anti-Rh agglutinins and subsequent exposures may result in a transfusion reaction.
- Hemolytic disease of the newborn is a threat when the mother is Rh- and the developing fetus is Rh+. If the mother has previously developed anti-Rh agglutinins, they may cross the placenta and enter the fetal blood causing agglutination and hemolysis.
- If hemolytic disease of the newborn develops, the fetal blood is temporarily replaced with Rh- blood.

☞ **Answers to Review Questions**

1. Volume of blood = 5 liters.
   pH = 7.35 to 7.45.
2. Oxygen, carbon dioxide, nitrogenous wastes, nutrients, hormones.
3. Regulation and protection.
4. 55% plasma, 45% formed elements.
5. Water.
6. Albumins: maintain osmotic pressure
   Globulins: lipid transport and immunity
   Fibrinogen: formation of blood clots
7. The three categories of formed elements are erythrocytes (red blood cells), leukocytes (white blood cells), and thrombocytes (platelets). These form from hemocytoblasts by a process of hemopoiesis.
8. Erythrocytes are anucleate, biconcave disks that contain hemoglobin and are about $7.5 \mu m$ in diameter. There are 4.5 to 6 million RBCs/mm^3 of blood
9. The primary function of erythrocytes is the transport of oxygen by the hemoglobin molecule.
10. Erythrocyte production is regulated by a negative feedback mechanism that uses the hormone erythropoietin. When blood oxygen is low, the kidneys produce renal erythropoietic factor, which activates erythropoietin, which stimulates erythropoiesis. Iron, vitamin B$_{12}$, and folic acid are necessary for erythrocyte production.
11. Erythrocytes live for about 120 days then are destroyed by the liver and spleen. The iron and protein portions of hemoglobin are reused by the body. The pigment portion of the molecule is converted to bilirubin and excreted in the bile.
12. Leukocytes have a nucleus, but do not have hemoglobin.
13. Granulocytes:
    Neutrophils--phagocytosis
    Eosinophils--counteract histamine
    Basophils--secrete histamine and heparin
    Agranulocytes:
    Lymphocytes--immunity
    Monocytes--phagocytosis
    Most numerous: neutrophils
    Least numerous: basophils
14. Thrombocytes are called platelets. They are fragments of megakaryocytes and function in blood clotting. The normal number ranges from 250,000 to 500,000 platelets/mm^3 of blood.
15. Hemostasis is the stoppage of bleeding. The first process is **vascular constriction** which reduces the flow of blood. In the second process, **platelet plug formation**, the platelets stick to the connective tissue around the opening and to each other to form a plug that obstructs the opening to reduce blood loss. The third and most effective process is the **formation of a blood clot**, or coagulation.
16. Damaged tissues →Prothrombin activator
    Prothrombin → Thrombin
    Fibrinogen → Fibrin
17. Calcium and vitamin K are necessary.
18. It is dissolved by fibrinolysis.

19. Type A has A agglutinogens and anti-B agglutinins; Type B has B agglutinogens and anti-A agglutinins; Type AB has both A and B agglutinogens and neither agglutinin; Type O has neither agglutinogen and both anti-A and anti-B agglutinins.

20. The agglutinins of the recipient react with the agglutinogens of the donor.

21. Type O is called the universal donor because it has no agglutinogens. Type AB is called the universal recipient because it has no agglutinins to react with a donor's agglutinogens.

22. Rh+ individuals have Rh agglutinogens on the RBCs; Rh− individuals do not.

23. When an Rh− individual is first exposed to Rh+ blood, the Rh− individual develops anti-Rh agglutinins. If there is a second exposure the anti-Rh agglutinins react with the Rh+ agglutinogens of the donor.

24. Hemolytic disease of the newborn is a problem that develops in the fetus when the mother is Rh− and the fetus is Rh+. If the mother previously has been exposed to Rh+ blood, either through transfusion or pregnancy, she develops anti-Rh agglutinins. These agglutinins cross the placenta and react with the Rh+ agglutinogens in the fetal blood. This causes agglutination and hemolysis in the fetal blood, which may obstruct blood vessels, cause kidney damage, and brain damage.

25. Hemolytic disease of the newborn may be treated by giving the newborn a complete blood transfusion of Rh− blood. This blood will be replaced slowly as the baby's hemopoietic mechanisms make new cells. The problem may be prevented by treating the mother with RhoGAM, an anti-Rh gamma globulin, during pregnancy. This inactivates any Rh+ agglutinogens that may enter the mother's blood and prevents the formation of anti-Rh agglutinins.

☞ **Answers to Learning Exercises**

Functions and Characteristics of the Blood (Objectives 1 and 2)
1. 7.35 to 7.45
    5 liters
    Gases
    Nutrients
    Waste products
    Body temperature
    Fluid and electrolyte balance
    pH

Composition of the Blood (Objectives 3-8)
1. Blood is **55%** plasma and **45%** formed elements.
2. Albumin - Osmotic pressure.
    36% - Lipid transport/immune reactions.
    Fibrinogen - 4%.
3. Amino acids, urea, uric acid.
    Amino acids, glucose, fatty acids.
    Na^+, K^+, Ca^{++}, Cl^-, HCO_3^-, PO_4^{-3}
4. Erythrocytes
    Leukocytes
    Thrombocytes
5. E (basophil)
    B (eosinophil)
    D (erythrocyte)
    F (lymphocyte)
    C (monocyte)
    A (neutrophil)
6. Erythrocytes
    Reticulocytes
    4.5 to 6 million
    Hemoglobin
    Diapedesis
    5,000 to 9,000
    Neutrophil
    Eosinophil
    Basophil
    Monocyte
    Lymphocyte
    Neutrophil
    Monocyte
    Neutrophil
    Monocyte
    Basophil
    Eosinophil
    Blood clotting
    250,000 to 500,000
    Megakaryocyte
7. Hemocytoblast
    Erythropoietin
    Iron
    Vitamin B_{12}
    Folic acid
    Intrinsic factor
    120 days or 4 months
    Spleen
    Liver
    Bilirubin

Hemostasis (Objectives 9-11)
1. Vascular constriction
    Platelet plug formation
    Vitamin K
    Calcium
    Fibrinolysis

2. Substance A = Prothrombin activator
   Substance B = Prothrombin
   Substance C = Thrombin
   Substance D = Fibrinogen

Blood Typing and Transfusion (Objectives 12-15)

1. A: anti-B: A and O: A and AB
   B: anti-A: B and O: B and AB
   AB: A and B: A, B, O, AB: AB
   O: anti-A and anti-B: O: A, B, AB, O

2. Rh+ (D)
   None
   None
   anti-Rh (anti-D)

3. The Rh− builds up antibodies (agglutinins) against Rh+
   Agglutination
   There are no Rh agglutinogens in the donor blood.
   Maternal type: Rh−   Fetal type: Rh+

☞ **Answers to Chapter Self-Quiz**

1. b
2. c
3. Hemopoiesis
   Hemocytoblast
   Oxyhemoglobin
   Neutrophil
   Erythropoietin
   Diapedesis
   Lymphocyte
4. T (fragments of large cells)
   E (contain hemoglobin)
   L (may have granules...)
   L (some are phagocytic)
   E (biconcave discs)
   T (platelets)
   E (transport oxygen)
   T (primary function is blood clotting)
   L (function in prevention of disease)
   E (normal number is about 5 million/mm^3
5. 4 (prothrombin is converted to thrombin)
   1 (smooth muscle in vessel walls contracts)
   3 (formation of prothrombin activator)
   5 (fibrinogen is converted to fibrin)
   2 (collagen attracts platelets to form plug)
6. B, O (has anti-A agglutinins)
   B, O (can be given to type B individuals)
   O   (has no agglutinogens)
   AB   (universal recipient)
   AB   (has no agglutinins)
   O   (can be given to type O individuals)
   A, O (can donate to type A individuals)
   B, AB (has type B agglutinogens)

   O   (universal donor)
   A, B, O (reacts with type AB donor blood

7. d
8. False
   True
   True
   True
   False

☞ **Answers to Terminology Exercises**

Hemolysis
Thrombocytopenia
Hemopoiesis
Basophilic
Fibrinolysis

Clumping together
A substance that inhibits clotting
A plasma protein that forms fibers
White cell
Clotting cell

E (substance before thrombin)
A (globe-shaped protein in blood)
D (condition of too many cells in blood)
B (flowing of blood)
C (cell with large nucleus)

☞ **Answers to Fun and Games**

PLASMA
FORMED ELEMENT
ERYTHROCYTE
GLOBULIN
ERYTHROPOIETIN
AGGLUTINATION
LEUKOCYTE
HISTAMINE
FIBRINOGEN
POLYCYTHEMIA
AGGLUTINOGEN
THROMBIN
LYMPHOCYTE
DIAPEDESIS
MACROPHAGE
EOSINOPHIL
AGGLUTININ
PROTHROMBIN
NEUTROPHIL
MAST CELL
PLATELET
BASOPHIL
BILIRUBIN
MEGAKARYOCYTE
GRANULOCYTE

HEMATOCRIT
THROMBOCYTE
ALBUMIN
AGRANULOCYTE
GAMMA GLOBULIN

☞   **Quiz/Test Questions**

Note: There are fifty multiple-choice questions for this chapter in the computerized test bank.

Name the following:

1. Component that makes up about 55 percent of blood volume.
   Answer: plasma.

2. Most abundant plasma protein.
   Answer: albumin.

3. Class of plasma proteins that function in transporting lipids and in immunity.
   Answer: globulins.

4. Stem cell or precursor cell that gives rise to all blood cell types.
   Answer: hemocytoblast.

5. Process of forming blood cells.
   Answer: hemopoiesis.

6. Most numerous phagocytic leukocyte.
   Answer: neutrophil.

7. Two agranulocytes in the blood.
   Answer: monocytes and lymphocytes.

8. Largest phagocytic blood cell.
   Answer: monocyte.

9. Molecule in erythrocytes that combines with oxygen.
   Answer: hemoglobin.

10. Formed element that functions in blood clotting.
    Answer: thrombocyte or platelet.

11. The insoluble threads that form the foundation of a blood clot.
    Answer: fibrin.

12. Blood type proteins on the surface of RBCs.
    Answer: agglutinogens.

13. Two ABO blood types that theoretically can be given to a type A individual.
    Answer: type A and type O.

14. Components in recipient's blood that react with mismatched donor's blood.
    Answer: agglutinins.

15. Fetal Rh type that may develop hemolytic disease of the newborn.
    Answer: Rh+.

True/False Questions:

1. Blood normally has a neutral pH of 7.
   Answer: False; normal blood pH ranges from 7.35 to 7.45, slightly alkaline.

2. Serum is another name for blood plasma.
   Answer: False; serum is the liquid that remains after a clot forms, it has the fibrinogen removed.

3. The most numerous formed element in the blood is the neutrophil.
   Answer: False; the most numerous formed element is the erythrocyte.

4. Granulocytes have red-staining granules in the cytoplasm.
   Answer: False; only the eosinophils have red-staining granules.

5. Iron, vitamin B_{12}, and folic acid are necessary for RBC production.
   Answer: True.

6. Immature RBCs that may be found in the blood are called reticulocytes.
   Answer: True.

7. Excessive hemolysis of RBCs may lead to an increased reticulocyte count.
   Answer: True.

8. Sodium and vitamin B_{12} are essential for the formation of a blood clot.
   Answer: False; calcium and vitamin K are essential for the formation of a blood clot.

9. Theoretically it is appropriate to give O− blood to an AB+ recipient.
   Answer: True, the recipient has no agglutinins to react with donor blood.

10. All babies born to untreated Rh− mothers are in danger of developing hemolytic disease of the newborn.
    Answer: False; only the Rh+ babies, the Rh− babies have no problem.

12 Heart

☞ **Key Terms/Concepts**

Atrioventricular valve Valve between an atrium and a ventricle in the heart.

Atrium Thin walled chambers of the heart that receive blood from veins.

Cardiac cycle A complete heartbeat consisting of contraction and relaxation of both atria and both ventricles.

Cardiac output The volume pumped from one ventricle in one minute; usually measured from the left ventricle.

Conduction myofibers Cardiac muscle cells specialized for conducting action potentials to the myocardium; part of the conduction system of the heart; also called Purkinje fibers.

Diastole Relaxation phase of the cardiac cycle; opposite of systole.

Electrocardiogram A graphic recording of the electrical changes that occur during a cardiac cycle.

Endocardium The thin, smooth, inner lining of each chamber of the heart.

Epicardium The outer layer of the heart wall; the visceral pericardium.

Myocardium Middle layer of the heart wall, composed of cardiac muscle tissue.

Pericardial cavity Potential space between the parietal pericardium and visceral pericardium that contains a small amount of serous fluid for lubrication.

Pericardium Membrane that surrounds the heart; usually refers to the pericardial sac.

Semilunar valve Valve between a ventricle of the heart and the vessel carries blood away from the ventricle; also pertains to the valves in veins.

Starling's law of the heart Principle that the more cardiac muscle fibers are stretched, the greater the contraction strength of the heart.

Stroke volume The volume of blood ejected from one ventricle during one contraction; normally about 70 ml.

Systole Contraction phase of the cardiac cycle; opposite of diastole.

Ventricle Chamber of the heart that receives blood from an atrium and pumps the blood away from the heart.

☞ **Chapter Objectives**

Upon completion of this chapter the student should be able to:

1. Describe the size and location of the heart.
2. Describe the pericardium and pericardial cavity.
3. Identify the layers of the heart wall and state the type of tissue in each layer.
4. Discuss the structure and function of each of the four chambers of the heart.
5. Identify the valves associated with the heart and describe their location, structure, and function.
6. Label a diagram of the heart, identifying the chambers, valves, and associated vessels.
7. Trace the pathway of blood flow through the heart, including chambers, valves, and pulmonary circulation.
8. Identify the major vessels that supply blood to the myocardium and return the deoxygenated blood to the right atrium.
9. Describe the components and function of the conduction system of the heart.
10. Correlate the deflections on an ECG with the electrical events in the conduction system.
11. Define systole and diastole.
12. Summarize the events of a complete cardiac cycle.
13. Correlate the heart sounds heard with a stethoscope with the events of the cardiac cycle.
14. Explain what is meant by cardiac output and describe the factors that affect it.
15. Describe the pathway be which the central nervous system regulates heart rate.

☞ **Chapter Outline/Summary**

Overview of the Heart (Objectives 1 and 2)
Form, size, and location of the heart (Objective 1)
● The heart is located in the middle mediastinum between the second and sixth ribs.
● The apex points downward and to the left so that 2/3 of the mass is on the left side.
● The heart is about the size of a closed fist.
Coverings of the heart (Objective 2)
● The heart is enclosed in a double layered pericardial sac. The outer layer is fibrous connective tissue. The inner layer is parietal serous membrane.

- The visceral layer of the serous membrane forms the surface of the heart and is called the epicardium.
- The space between the parietal and visceral layers of the serous membrane is the pericardial cavity.

Structure of the Heart (Objectives 3-8)
Layers of the heart wall (Objective 3)

- The outermost layer of the heart wall is the visceral layer of the serous pericardium and is called the epicardium.
- The middle layer of the heart wall is the cardiac muscle tissue. It is the thickest layer and is called the myocardium.
- The innermost layer is simple squamous epithelium and is called the endocardium.

Chambers of the heart (Objective 4)

- The right atrium is a thin walled chamber that receives deoxygenated blood from the superior vena cava, inferior vena cava, and coronary sinus.
- The right ventricle receives blood from the right atrium and pumps it out to the lungs to receive oxygen. It has a thick myocardium.
- The left atrium is a thin walled chamber that receives oxygenated blood from the lungs through the pulmonary veins.
- The left ventricle has the thickest myocardium. It receives the oxygenated blood from the left atrium and pumps it out to systemic circulation.

Valves of the heart (Objective 5)

- There are two types of valves associated with the heart--atrioventricular valves and semilunar valves.
- Atrioventricular valves are located between the atria and ventricles. They prevent blood from flowing back into the atria when the ventricles contract.
- Semilunar valves are located at the exits from the ventricles and prevent blood from flowing back into the ventricles when the ventricles relax.
- The AV valve on the right side is the tricuspid valve. On the left, it is the bicuspid, or mitral, valve.
- The pulmonary semilunar valve is located at the exit of the right ventricle. The aortic semilunar valve is at the exit of the left ventricle.

Pathway of blood through the heart (Objective 7)

- Deoxygenated blood enters the right atrium through the superior vena cava, inferior vena cava, and coronary sinus.
- Atrial systole lasts for 0.1 sec. followed by ventricular systole for 0.3 sec.

- From the right atrium, the blood goes through the tricuspid valve into the right ventricle, then is pumped through the pulmonary semilunar valve into the pulmonary trunk, then pulmonary arteries, to the capillaries of the lungs.
- In the lung capillaries, the blood gives off CO_2, picks up O_2, then enters the pulmonary veins, and flows into the left atrium.
- Oxygenated blood in the left atrium goes through the bicuspid valve into the left ventricle, then is pumped through the aortic semilunar valve into the ascending aorta to enter systemic circulation.

Blood supply to the myocardium (Objective 8)

- The right and left coronary arteries, branches of the ascending aorta, supply blood to the myocardium in the wall of the heart.
- Blood from the capillaries in the myocardium enters the cardiac veins, which drain into the coronary sinus. From there, it enters the right atrium.

Physiology of the Heart (Objectives 9-15)
Conduction system (Objectives 9 and 10)

- Components of the conduction system (Objective 9)
  - The conduction system of the heart consists of specialized cardiac muscle cells that act in a manner similar to neural tissue. The conduction system coordinates the contraction and relaxation of the heart chambers.
  - The SA node has the fastest rate of depolarization, therefore it is the pacemaker in the conduction system.
  - Other components in the conduction system are the AV node, AV bundle, bundle branches, and conduction myofibers.
- Electrocardiogram (Objective 10)
  - An electrocardiogram is a recording of the electrical activity of the heart.
  - The P wave is produced by depolarization of the atrial myocardium.
  - The QRS wave is produced by depolarization of the ventricular myocardium and repolarization of the atria.
  - The T wave is due to repolarization of the ventricles.

Cardiac cycle (Objectives 11 and 12)

- Systole is the contraction phase of the cardiac cycle and diastole is the relaxation phase.
- At a normal heart rate, one complete cardiac cycle lasts for 0.8 sec.
- All chambers are in diastole at the same time for 0.4 sec.

- Most ventricular filling occurs while all chambers are relaxed.

Heart sounds (Objective 13)

- Heart sounds are due to vibrations in the blood caused by the valves closing.
- The first heart sound is caused by closure of the AV valves.
- The second heart sound is caused by closure of the semilunar valves.

Cardiac output (Objectives 14 - 15)

- Cardiac output equals stroke volume times heart rate. Anything that affects either component affects the output.
- Stroke volume (Objective 14)
  - Stroke volume is influenced by end diastolic volume and contraction strength.
  - End diastolic volume depends on venous return.
  - Contraction strength depends on end diastolic volume and stimulation by the autonomic nervous system.
- Heart rate (Objective 15)
  - The cardiac center in the medulla oblongata has both sympathetic and parasympathetic components that adjust the heart rate to meet the changing needs of the body.
  - Peripheral baroreceptors and chemoreceptors send impulse to the cardiac center where appropriate responses adjust heart rate.
  - Emotions and body temperature also affect heart rate. These effects are usually coordinated through the cardiac center.

☞ **Answers to Review Questions**

1. Between the second and sixth ribs (or fifth intercostal space). The apex is at the fifth intercostal space and the base is at the level of the second rib.
2. The outer fibrous layer of fibrous connective tissue and the inner layer, which is the parietal layer of serous membrane.
3. The pericardial cavity is between the visceral and parietal layers of the serous membrane.
4. Outer layer: epicardium, which is the visceral layer of serous membrane.
   Middle layer: myocardium, which is cardiac muscle.
   Inner layer: endocardium, which is simple squamous epithelium.
5. The atria are thin-walled chambers and the ventricles have thick walls. The difference is in the thickness of the myocardium.

6. The thin-walled atria are receiving chambers and do not have to generate much force in contraction. The ventricles are pumping chambers and need to generate considerable force, thus they have thick myocardium.
7. Each AV valve consists of a fibrous ring and double folds of endothelium that form cusps. The cusps are anchored to papillary muscles by chordae tendineae. Each SL valve consists of three cuplike cusps of endothelium.
8. The AV valves prevent the backward flow of blood when the ventricles contract and blood is under pressure. The SL valves open under the force of pressure when the ventricles contract and close when the pressure decreases during ventricular relaxation.
9. See Figure 12-3 in the textbook.
10. Right atrium
    Tricuspid valve
    Right ventricle
    Pulmonary semilunar valve
    Pulmonary trunk
    Pulmonary arteries
    Lungs
    Pulmonary veins
    Left atrium
    Bicuspid (mitral) valve
    Left ventricle
    Aortic semilunar valve
    Ascending aorta
11. The myocardium receives the oxygen that is required for contraction through a system of coronary arteries. The right and left coronary arteries branch from the ascending aorta.
12. Sinoatrial (SA) node
    Atrioventricular (AV) node
    Atrioventricular bundle (of His)
    Right and left bundle branches
    Conduction myofibers (Purkinje fibers)
    The conduction system coordinates the contraction and relaxation of the heart chambers. The intrinsic rhythm of depolarization of cardiac muscle is no fast enough to be compatible with life.
13. The P wave corresponds to atrial depolarization as the impulse spreads from the SA node to the AV node. The QRS wave corresponds to ventricular depolarization as the impulse spreads throughout the conduction myofibrils. The T wave corresponds to ventricular repolarization.
14. Systole is the contraction phase, diastole is the relaxation phase of the cardiac cycle.

15. Atrial systole: both atria contract simultaneously for 0.1 second; AV valves are open and ventricles are in diastole.

    Atrial diastole: both atria relax simultaneously, which lasts for 0.7 second; ventricular systole occurs during the first portion of this time.

    Ventricular systole: both ventricles contract simultaneously for 0.3 seconds right after atrial systole; AV valves close, SL valves open.

    Ventricular diastole: both ventricles relax simultaneously for 0.5 seconds; atrial systole occurs during the final second of ventricular diastole; SL valves close and AV valves open.

16. The first heart sound is due to turbulence in the blood as a result of AV valves closing. The second heart sound is due to turbulence in the blood as a result of SL valves closing.

17. Cardiac output = Heart rate X Stroke volume

18. End diastolic volume is directly related to stroke volume because the more blood there is present in the ventricles, the more can be ejected from the heart. Increased end diastolic volume also increases the strength of contraction, by Starling's law of the heart, which also increases stroke volume. As stroke volume increases, cardiac output increases.

19. The cardiac center in the medulla oblongata regulates the rate of the SA node. This action occurs through sympathetic and parasympathetic components.

☞ **Answers to Learning Exercises**

Overview of the Heart (Objectives 1 and 2)
1. The **apex** of the heart is directed inferiorly and to the left.
   About **2/3** of the heart mass is on the left side.
   The heart is located in the **middle** mediastinum, between the second and **sixth** ribs.
2. 4 (epicardium)
   3 (pericardial cavity)
   1 (fibrous pericardium)
   2 (parietal pericardium)

Structure of the Heart (Objective 3-8)
1. Myocardium
   Right atrium
   Tricuspid
   Endocardium
   Left atrium
   Fossa ovalis
   Right ventricle
   Trabeculae carneae
   Bicuspid (mitral)
   Aortic semilunar valve
   Coronary arteries
   Right atrium
2. M (right atrium)
   Q (right ventricle)
   F (left atrium)
   I (left ventricle)
   R (interventricular septum)
   L (superior vena cava)
   P (inferior vena cava)
   E (pulmonary trunk)
   C (ascending aorta)
   K (right brachiocephalic vein)
   J (brachiocephalic artery)
   G (aortic semilunar valve)
   N (pulmonary semilunar valve)
   O (tricuspid valve)
   H (bicuspid valve)
   A (left common carotid artery)
   B (left subclavian artery)
   D (left pulmonary artery)
3. 13 (aortic semilunar valve)
   14 (ascending aorta)
   11 (bicuspid valve)
   8 (capillaries of lungs)
   10 (left atrium)
   12 (left ventricle)
   5 (pulmonary semilunar valve)
   6 (pulmonary trunk)
   9 (pulmonary veins)
   7 (pulmonary arteries)
   2 (right atrium)
   4 (right ventricle)
   3 (tricuspid valve)
   1 (venae cavae)

Physiology of the Heart (Objectives 9-15)
1. Sinoatrial node
   0.8 second
   P wave
   Sinoatrial node
   Electrocardiogram
   QRS wave
   Systole
   Conduction myofibers
   T wave
   Diastole
   Atrioventricular bundle
   0.1 second
   sinoatrial node
   0.3 second

2. D (atria are in systole)
   D (atrioventricular valves open)
   S (atrioventricular valves close)
   S (semilunar valves open)
   D (semilunar valves close)
   S (blood is ejected from the heart)
   D (pressure in ventricles decreases)
   D (blood enters ventricles)
3. a. AV valves
   b. SL valves
4. False   plus should be times
   True
   False   decreases should be increases
   False   decreases should be increases
   True
   True
   False   decreases should be increases
   True
   True
   False   pons should be medulla oblongata
   False   decreases should be increases
   True

☞ Answers to Chapter Self-Quiz

1. d
2. False--The outermost layer of the heart wall is the epicardium.
3. The innermost layer of the heart wall is the endocardium, which is simple squamous epithelium. The middle layer is the myocardium, which is the cardiac muscle layer. The outermost layer is the epicardium, which is a serous membrane.
4. LA (receives blood from pulmonary veins)
   RV (pumps blood to the lungs)
   RA, RV (contains unoxygenated blood)
   RV, LV (contains papillary muscles)
   RA (receives blood from the IVC)
   LV (pumps blood through the aortic valve)
   LV (pumps blood into systemic circulation)
   RV (receives blood through the tricuspid)
   RA (coronary sinus opens into this)
   LA (bicuspid valve is at the exit)
5. Anterior descending (interventricular) artery
   Circumflex artery
6. False   (atrial systole...)
   True   (all chambers...)
   True   (an increase...)
   False   (most of the blood...)
   True   (the P wave...)
   False   (the first heart sound...)
   False   (the T wave...)
   True   (impulses are carried...)
   False   (ventricular diastole...)
   False   (blood is oxygenated...)

7. 75 beats/min X 65 ml/beat = 4875 ml/min
8. b

☞ Answers to Terminology Exercises

Atrium
Tachycardia
Tricuspid
Aorta
Semilunar

Enlarged heart
Narrowing of the mitral valve opening
Inflammation of a valve
Slow heart rate
Contraction

D (membrane around the heart)
A (disease condition of heart muscle)
E (surgical repair of a valve)
C (inflammation of heart lining)
B (recording of electrical activity of heart)

☞ Answers to Fun and Games

```
C                       T   S               V       B A S E
I R   E N D O C A R D I T I S           E       T
C     N       A   I   E               I         R
U     D   T   R   C       N           N         O
M Y O C A R D I U M   O   B I C U S P I D       K
F     C   C   I   S       S   R               I E
L A   H   O   S       P E R I C A R D I U M   A Y
E R   Y   M   I   S       D         S         S O
X     D   C   E   D       Y       S Y S T O L E
      I   A   G           C             O   U M
      U   R   A U S C U L T A T I O N   L   M E
S     M   D   L       R               E     E
E     I   I   Y   N   V   D   A   D
M I T R A L       O   E P I C A R D I U M
I     L           D   N   A   T   P
L                 E   T       E   P
U   A P E X       A O R T A   R
N   T             I       Y
A U R I C L E     C
R   I     C   V A L V U L I T I S       Q
    L U B B   G   E                     R
    M             F O S S A O V A L I S
```

WORD LIST

ECG	Systole	Bradycardia
QRS	Auricle	Pericardium
		Endocardium
Apex	Stenosis	Fossa ovalis
Base	Diastole	Tachycardia
Lubb	Bicuspid	
Dupp		Auscultation
	Ventricle	Cardiomegaly
Aorta	Tricuspid	Stroke volume
Veins	Semilunar	Endocarditis
Atrium	Epicardium	
SA node	Myocardium	
Mitral	Valvulitis	
Artery	Circumflex	

☞ Quiz/Test Questions

Note: There are fifty multiple-choice questions for this chapter in the computerized test bank.

Name the following:

1. Thickest layer of the heart wall.
   Answer: myocardium; cardiac muscle layer.

2. Chamber of the heart that receives oxygenated blood from the lungs.
   Answer: left atrium.

3. Chamber of the heart with the thickest wall.
   Answer: left ventricle.

4. Atrioventricular valve that is associated with the same chamber as the pulmonary semilunar valve.
   Answer: tricuspid valve.

5. Projections of myocardium that anchor atrioventricular valve cusps.
   Answer: papillary muscles.

6. Chamber that receives venous blood from the superior vena cava.
   Answer: right atrium.

7. Vessel that branches to form the circumflex artery.
   Answer: left coronary artery.

8. Part of the conduction that is called the pacemaker.
   Answer: sinoatrial (SA) node.

9. Contraction phase of the cardiac cycle.
   Answer: systole.

10. Portion of cardiac cycle that lasts about 0.3 seconds in normal resting cycle.
    Answer: ventricular systole.

11. Region of an ECG that represents ventricular depolarization.
    Answer: QRS complex.

12. Term given to amount of blood ejected from a ventricle in one contraction.
    Answer: stroke volume.

13. Term given to amount of blood in ventricle at beginning of contraction.
    Answer: end diastolic volume.

14. Location of the cardiac center that alters heart rate.
    Answer: medulla oblongata.

15. Portion of the autonomic nervous system that generally increases heart rate and contraction strength.
    Answer: sympathetic division.

True/False Questions:

1. The fibrous pericardial sac is lined with a serous membrane.
   Answer: True; the parietal layer of serous pericardium.

2. The right ventricle makes up most of the apex of the heart.
   Answer: False; the left ventricle makes up the apex.

3. The superior vena cava enters the heart on the right side.
   Answer: True.

4. The base of the heart is more superior than the apex.
   Answer: True.

5. The first heart sound is due to turbulence in the blood resulting from closure of the semilunar valves.
   Answer: False; from closure of atrioventricular valves.

6. All chambers of the heart are in diastole at the same time for about half the duration of a normal cardiac cycle.
   Answer: True.

7. Most of the blood that is ejected from the ventricles enters the ventricles during atrial systole.
   Answer: False; 70 percent of ventricular filling occurs during atrial diastole, before the atria contract.

8. Increased venous return means there is greater end diastolic volume, which results in more forceful contraction of the ventricular myocardium.
   Answer: True; Starling's law.

9. Parasympathetic stimulation generally increases cardiac output.
   Answer: False; sympathetic stimulation increases cardiac output.

10. Baroreceptors and chemoreceptors send signals to the cardiac center in the pons which initiates sympathetic impulses to increase heart rate.
    Answer: False; the cardiac center is in the medulla oblongata.

13 Blood Vessels

Key Terms/Concepts

Artery A blood vessel that carries blood away from the heart.

Blood pressure Pressure exerted by the blood against the vessel walls; usually refers to arterial blood pressure.

Capillary Microscopic blood vessel between an arteriole and a venule, where gaseous exchange takes place.

Diastolic pressure Blood pressure in the arteries during relaxation of the ventricles.

Korotkoff sounds The sounds heard in the stethoscope while taking blood pressure.

Metarteriole A microscopic vessel that directly connects an arteriole to a venule without an intervening capillary network; an arteriovenous shunt.

Peripheral resistance Opposition to blood flow cause by friction of the blood vessel walls.

Pulmonary vessels Blood vessels that transport blood from the heart to the lungs and then return it to the left atrium.

Pulse Expansion and recoil of arteries caused by contraction and relaxation of the heart.

Pulse pressure Difference between systolic and diastolic pressures.

Systemic vessels Blood vessels that transport blood from the heart to all parts of the body and back to the right atrium.

Systolic pressure Blood pressure in the arteries during contraction of the ventricles.

Vasa vasorum Small blood vessels that supply nutrients to the tissues in the walls of the large blood vessels.

Vasoconstriction A narrowing of blood vessels; decrease in the size of the lumen of blood vessels.

Vasodilation An enlarging of blood vessels; increase in the size of the lumen of blood vessels.

Vein A blood vessel that carries blood toward the heart.

Chapter Objectives

Upon completion of this chapter the student should be able to:

- Describe the structure and function of arteries, capillaries, and veins.
- Discuss how oxygen, carbon dioxide, glucose, and water move across capillary walls.
- Discuss three factors that affect blood flow in arteries.
- State three actions that provide pressure gradients for blood flow in veins.
- Identify at least six commonly used pulse points.
- Distinguish between systolic pressure, diastolic pressure, and pulse pressure. State normal values for each.
- Discuss four primary factors that affect blood pressure.
- Explain the role of baroreceptors and chemoreceptors in the regulation of blood pressure.
- Describe the renin/angiotensin/aldosterone mechanism for blood pressure regulation.
- Trace blood through the pulmonary circuit from the right atrium to the left atrium.
- Identify the major systemic arteries.
- Describe the blood supply to the brain.
- Identify the major systemic veins.
- Describe five features of fetal circulation that make it different from adult circulation.

Chapter Outline/Summary

Classification and Structure of Blood Vessels (Objective 1)
Arteries
- Arteries carry blood away from the heart.
- The wall of an artery consists of the tunica intima (simple squamous epithelium), tunica media (smooth muscle), and tunica externa (connective tissue).
Capillaries
- Capillaries, form the connection between arteries and veins.
- Capillary walls are simple squamous epithelium.
- Capillaries function in the exchange of materials between the blood and the tissue cells.

- The number of capillaries in a tissue depends on the metabolic activity of the tissue. Metabolically active tissues have an abundance of capillaries; less active tissues have fewer capillaries.
- Small arterioles and precapillary sphincters regulate blood flow into the capillaries.

Veins

- Veins carry blood toward the heart.
- The walls of veins have the same three layers as the arteries, but the layers are thinner.
- Veins have valves to prevent the backflow of blood.

Physiology of Circulation (Objectives 2-9)

Role of the capillaries (Objective 2)

- Capillaries have a vital role in the exchange of gases, nutrients, and metabolic waste products between the blood and tissue cells.
- Solutes such as oxygen, carbon dioxide, and glucose move across the capillary wall by diffusion.
- A combination of hydrostatic pressure and osmotic pressure determines fluid movement across the capillary wall.
- Fluid moves out of the capillary at the arteriole end and returns at the venous end.

Blood flow (Objectives 3 and 4)

- Relationship of flow to pressure (Objective 3)
  - Blood flows from a higher pressure area to a lower pressure area; it flows in the same direction as the pressure gradient.
  - Pressure is lowest as the venae cavae enter the right atrium. Pressure in the right atrium is called the central venous pressure.
- Velocity of blood flow (Objective 3)
  - Velocity of blood flow varies inversely with the total cross-sectional area of the blood vessels. As the area increases, the velocity decreases.
  - Since capillaries are so numerous, they have the greatest cross-sectional area and the slowest blood flow.
- Relationship of flow to resistance (Objective 3)
  - Resistance is a force that opposes blood flow.
  - As resistance increases, blood flow decreases.
  - The autonomic nervous system regulates blood flow by changing the resistance of the vessels through vasoconstriction and vasodilation.

- Venous blood flow (Objective 4)
  - Very little pressure from ventricular contraction remains by the time the blood reaches the veins.
  - Venous blood flow depends on skeletal muscle action, respiratory movements, and contraction of smooth muscle in venous walls.

Pulse and blood pressure (Objectives 5-9)

- Meaning of pulse (Objective 5)
  - Pulse refers to the rhythmic expansion of an artery that is caused by ejection of blood from the ventricle.
  - Pulse can be felt where an artery is close to the surface and rests on something firm. These locations are called pulse points.
- Blood pressure and its determination (Objective 6)
  - Systolic pressure is the pressure in the arteries during ventricular contraction (systole). Normal systolic pressure is about 120 mm Hg.
  - Diastolic pressure is the pressure in the arteries during ventricular relaxation (diastole). Normal diastolic pressure is about 80 mm Hg.
  - Pulse pressure is the difference between systolic pressure and diastolic pressure. This is usually about 40 mm Hg.
  - A sphygmomanometer is used to measure blood pressure. The brachial artery is a common site for measuring blood pressure.
- Factors that affect blood pressure (Objective 7)
  - Cardiac output
  - Blood volume
  - Peripheral resistance
  - Viscosity of the blood
- Regulation of blood pressure (Objectives 8 and 9)
  - Baroreceptors in the aortic arch and carotid sinus detect stretching in the vessel walls when blood pressure increases. Signals are relayed back to the heart and blood vessels that reduce the pressure.
  - Baroreceptors are important in short term blood pressure regulation.
  - Chemoreceptors detect carbon dioxide, hydrogen ion, and oxygen concentrations. This is significant only in emergency situations.

○ Antidiuretic hormone plays a role in regulating blood volume, which has an affect on blood pressure.

○ In response to decreases in blood pressure, the kidneys secrete renin, which stimulates the production of angiotensin. Angiotensin causes vasoconstriction and promotes the release of aldosterone. Both actions result in increased blood pressure.

Circulatory Pathways (Objectives 10-14)

Pulmonary circuit (Objective 10)

● Pulmonary circulation transports oxygen poor blood from the right ventricle to the lungs where the blood picks up a new blood supply then it returns the oxygen rich blood to the left atrium.

● Figure 13-10 shows the pathway of pulmonary circulation.

Systemic circuit (Objectives 11-13)

● Major systemic arteries (Objectives 11 and 12)

○ Systemic arteries carry oxygenated blood from the left ventricle to the capillaries in the tissues of the body.

○ Figures 13-11, 13-12, and 13-13 illustrate the major systemic arteries.

○ The principal blood supply to the brain is through the internal carotid arteries and the vertebral arteries. Branches of these vessels form the circle of Willis at the base of the brain.

● Major systemic veins (Objective 13)

○ Systemic veins carry oxygen poor, carbon dioxide laden blood from the tissues to the right atrium of the heart.

○ Figures 13-15, 13-16, and 13-17 illustrate the major systemic veins.

Blood supply to the fetus (Objective 14)

● There are some differences in the circulatory pathways of the fetus because the lungs, gastrointestinal tract, and kidneys are not functioning.

● The two umbilical arteries carry fetal blood to the placenta; the umbilical vein carries blood from the placenta to the fetus; the placenta functions in the exchange of gases and nutrients between the maternal and fetal blood; the ductus venosus allows blood to bypass the immature fetal liver; the foramen ovale and ductus arteriosus permit blood to bypass the fetal lungs.

☞ **Answers to Review Questions**

1. Tunica externa, tunica media, tunica intima.
2. Veins have thinner walls than arteries and veins have valves but arteries do not.
3. Diffusion.
4. Hydrostatic (blood) pressure and osmotic pressure.
5. Rate of flow decreases as blood moves into the more numerous smaller vessels.
6. Blood tends to pool in the legs and feet when standing a long time because there is a lack of skeletal muscle contraction to "pump" the blood upward.
7. Radial artery.
8. Pulse pressure $= 115 - 78$ mm Hg $= 37$ mm Hg
9. Systolic pressure increases because the vessel cannot expand to accommodate the blood. This results in an increased pulse pressure.
10. The vasomotor and cardiac centers in the medulla oblongata respond to signals from the baroreceptors. The centers respond by sending out signals to decrease heart rate and dilate the blood vessels to decrease blood pressure.
11. Renin causes the production of angiotensin, which is a powerful vasoconstrictor and increases blood pressure. Angiotensin also promotes the secretion of aldosterone from the adrenal cortex. This causes sodium and water retention and increases blood pressure.
12. Tricuspid valve and pulmonary semilunar valve.
13. Brachiocephalic artery.
14. Vertebral artery.
15. Right and left brachiocephalic veins.
16. Abdominal aorta → celiac trunk → splenic artery → capillaries of the spleen → splenic vein → hepatic portal vein → sinusoids of liver → hepatic vein → inferior vena cava.
17. The foramen ovale, which permits blood to pass from the right atrium directly into the left atrium, and the ductus arteriosus, which permits blood to pass from the pulmonary trunk into the descending aorta.

☞ **Answers to Learning Exercises**

Classification and Structure of Blood Vessels (Objective 1)

1. A (carries blood away...)
   C (functions in the exchange...)
   C (wall consists of simple...)
   V (carries blood toward heart)
   V (has valves)
   A (has three relatively thick...)

V (has three relatively thin...)
A (may have vasa vasorum)
2. a) connective tissue - tunica externa
   b) simple squamous epithelium - tunica intima
   c) smooth muscle - tunica media

Physiology of Circulation (Objectives 2-9)
1. Diffusion
   Diffusion
   Osmosis and
   Filtration
2. V (osmotic pressure is greater...)
   A (hydrostatic pressure is greater...)
   A (net movement of fluid out...)
   V (net movement of fluid into...)
3. Aorta
   Capillary
   Vein
   Capillary
   Artery
   Blood pressure
   Skeletal muscle contraction
   Respiratory movements
   Smooth muscle vasoconstriction
   Systolic
   Diastolic
   Sphygmomanometer
   Pulse pressure
   Korotkoff sounds
   Relaxes
4. A) Common carotid a.
   B) Brachial a.
   C) Radial a.
   D) Dorsalis pedis a.
   E) Temporal a.
   F) Facial a.
   G) Axillary a.
   H) Femoral a.
5. Medulla oblongata
   Cardiac center
   Vasomotor center
   Baroreceptors (pressoreceptors)
   Chemoreceptors
   Renin
   Vasoconstriction
   Increase aldosterone
   Aldosterone
6. Checks (✓) for Increase:
      A marked increase in blood volume
      An increase in heart rate
      Polycythemia
      Sympathetic stimulation of arterioles
      Production of angiotensin
   Checks (✓) for Decrease
      Loss of body fluids
      Vasodilation
      Slow heart beat

Circulatory Pathways (Objectives 10-14)
1. Right atrium → **tricuspid valve** → right ventricle → **pulmonary semilunar valve** → **pulmonary trunk** → **pulmonary artery** → capillaries of lungs → **pulmonary veins** → left atrium.
2. A) Anterior cerebral a.
   B) Internal carotid a.
   C) Posterior communicating a.
   D) Basilar a.
   E) Middle cerebral a.
   F) Posterior cerebral a.
   G) Vertebral a.
3. Coronary a.
   Brachiocephalic a.
   Left subclavian a.
   Common carotid a.
   Vertebral a.
   Left subclavian a.
4. Brachial a.
   Radial a.
   Ulnar a.
   Common hepatic a.
   Left gastric a.
   Splenic a.
   Superior mesenteric a.
   Renal a.
   Inferior mesenteric a.
   Common iliac a.
   Internal iliac a.
   Femoral a.
   Popliteal a.
   Posterior tibial a.
   Anterior tibial a.
5. Inferior vena cava
   Superior vena cava
   Right and left brachiocephalic v.
   Internal jugular v.
   Basilic v.
   Cephalic v.
   Median cubital v.
   Brachial v.
   Azygos v.
   Hepatic v.
   Hepatic portal v.
   Splenic v.
   Superior mesenteric v.
   Great saphenous v.
   Common iliac v.
   Internal iliac v.
   Femoral v.
   Anterior tibial v.
   Posterior tibial v.
6. Umbilical artery
   Umbilical vein
   Ductus venosus

Foramen ovale

Ductus arteriosus

Placenta

7. Left ventricle → aortic semilunar valve → **ascending aorta** → aortic arch → **brachiocephalic artery** → **right common carotid artery** → **right internal carotid artery** → right middle cerebral artery.

8. Left gonad → left gonadal vein → **left renal vein** → **IVC** → right atrium → **tricuspid** valve → **right ventricle** → **pulmonary semilunar** valve → **pulmonary trunk** → **pulmonary artery** → capillaries of lungs → **pulmonary vein** → **left atrium** → **bicuspid** valve → **left ventricle** → **aortic semilunar** valve → **ascending aorta** → aortic arch → **left subclavian artery** → **left axillary artery** → **left brachial artery** → **left radial artery** → capillaries on lateral side of left forearm.

9. Descending aorta → **celiac trunk** → **splenic artery** → capillaries of the spleen → **splenic vein** → **hepatic portal vein** → sinusoids of liver → **hepatic vein** → **IVC** → right atrium.

☞ Answers to Chapter Self-Quiz

1. b
2. e
3. I (increase the blood pressure)
   D (increase the resistance)
   I (increase the cardiac output)
   I (vasodilation)
   I (increase in carbon dioxide conc.)
   D (contraction of precapillary sphincters)
   I (decrease in pH)
4. A (pressure when first Korotkoff...)
   C (normal is about 40 mm Hg)
   B (blood begins to flow freely)
   A (normally is about 120 mm Hg)
   B (normally is about 80 mm Hg)
5. I (increase in heart rate)
   D (decreased ADH)
   I (polycythemia)
   D (decrease in blood volume)
   D (vasodilation)
   I (decreased elasticity)
   D (stimulation of baroreceptors)
   I (epinephrine)
   I (fluid retention)
   I (increased angiotensin)
6. I (carries deoxygenated blood to lungs)
   B (branch from aortic arch...)
   D (formed from the bifurcation...)
   O (branch of the subclavian artery...)
   N (artery on medial side of forearm)

C (branch of aorta to liver and spleen)
M (branch of aorta to small intestine)
F (large artery in thigh)
A (artery to anterior portion of leg)
K (artery that supplies blood to kidney)

7. K (receives blood from venous sinuses)
   P (carries blood from placenta to liver)
   N (receives blood from head)
   B (deep vein in the arm)
   L (single vein in posterior knee region)
   D (superficial vein on lateral forearm)
   H (drains blood from liver to IVC)
   G (superficial vein of thigh and leg)
   I (vessel between SMV and liver)
   E (veins that join to form IVC)

☞ Answers to Terminology Exercises

Brachiocephalic
Arteriosclerosis
Phlebitis
Atheroma
Embolism

Study of vessels
Inflammation of an artery
Excision of a vein
Condition of swelling
Narrowing the opening of a blood vessel

E (crushing a blood vessel to stop hemorrhage)
C (surgical repair of a blood vessel)
B (disease of a blood vessel)
D (surgical excision of an artery)
A (tumor of a blood vessel)

☞ Answers to Fun and Games

```
· · L · · · L E · · · · · · · ·
· U M · · · N · · F T · · · · · ·
P · N · · I E V Y · A T R · · · R · ·
· E · A · · · · R · I · · · T A ·→
· · L · R · · · A · · · U · A · ·
· C · · Y S · · N · · · C I B M · · E ·
· I · · · E · · · M · U S P I D · V ·
· · R T · · M · · L · P S · · V L ·
· · · N · · I L U · · U · G A · A · ·
E R · · E · · N A R · · N L · V R · ·
V · I · · V · · V · · L U · · V · A · ·
· L · G H T · · A · Y · · · L E N · ·
· · A V · · · L · · R Y R · · E U · ·
· · · U D · · V · · · E A · A · F L I M
· R I M I · · · · E · T R · N · M T V E
T · · · T P · P U L · · · · · · · L E S
A T · · R S · · · M · · · · · P U · N C
· · H · · I U · · · · R U N K · · T · I
· · G I · · C · N · T · · · · I R · R T
· · · R · · · A R Y · · · C L E A O ·
```

BLOOD PATHWAY:
> Right atrium
> Tricuspid valve
> Right ventricle
> Pulmonary semilunar valve
> Pulmonary trunk
> Pulmonary artery
> Lungs
> Pulmonary vein
> Left atrium
> Bicuspid valve
> Left ventricle
> Aortic semilunar valve
> Aorta

☞ Quiz/Test Questions

Note: There are fifty multiple-choice questions for this chapter in the computerized test bank.

Name the following:

1. Smallest blood vessels.
   Answer: capillaries.

2. Vessels that carry blood away from the heart.
   Answer: arteries.

3. Two differences in structure between arteries and veins.
   Answer: veins have thinner walls and have valves.

4. Term for systolic pressure minus diastolic pressure.
   Answer: pulse pressure.

5. Two pressures that determine fluid movement across a capillary wall.
   Answer: hydrostatic (blood) pressure and osmotic pressure.

6. Pulse point in the neck.
   Answer: common carotid artery.

7. Sounds heard through a stethoscope as a result of blood flow.
   Answer: Korotkoff sounds.

8. Location of the vasomotor center in the CNS.
   Answer: medulla oblongata.

9. Enzyme from the kidney that has a role in blood pressure regulation.
   Answer: renin.

10. Branch of the aortic arch that supplies blood to the right side of the head and right arm.
    Answer: brachiocephalic artery.

11. Branch of the abdominal aorta that supplies blood to the liver and spleen.
    Answer: celiac trunk.

12. Vessels formed by bifurcation of the aorta.
    Answer: right and left common iliac arteries.

13. Vessel that carries blood from the digestive organs and spleen to the liver.
    Answer: hepatic portal vein.

14. Two veins that join to form the SVC.
    Answer: right and left brachiocephalic veins.

15. Vessel in fetal circulation that permits blood to bypass the fetal liver.
    Answer: ductus venosus.

True/False Questions:

1. Blood flow is slower in the capillaries of the finger than in the brachial vein of the arm.
   Answer: True; flow is slowest in the capillaries.

2. All of the fluid that leaves a capillary at the arterial end returns to the blood at the venous end.
   Answer: False; about 90 percent returns at venous end, the other 10 percent becomes interstitial fluid.

3. The artery in the forearm that is frequently used for taking pulse is the ulnar artery.
   Answer: False; it is the radial artery.

4. Diastolic pressure is usually greater than systolic pressure.
   Answer: False; systolic is greater than diastolic.

5. As peripheral resistance increases, blood flow and blood pressure decreases.
   Answer: False; as peripheral resistance increases, blood pressure also increases.

6. An increase in either blood volume or blood viscosity tends to increase blood pressure.
   Answer: True.

7. Decreases in blood pressure reduce the frequency of action potentials from the pressure receptors in the aortic arch and this results in increased heart rate and vasoconstriction.
   Answer: True.

8. Angiotensin causes vasoconstriction and the release of aldosterone from the adrenal cortex, both of which increase blood pressure.
   Answer: True.

9. Skeletal muscle action and respiratory movements create pressure gradients that facilitate flow of blood in the veins.
   Answer: True.

10. The renal veins drain into the hepatic portal vein before entering the inferior vena cava.
    Answer: False; renal veins drain directly into the inferior vena cava.

14 Lymphatic System and Immunity

☞ **Key Terms/Concepts**

Active immunity Immunity that is produced as the result of an encounter with an antigen and memory cells are produced.

Antibody Substance produced by the body that inactivates or destroys another substance that is introduced into the body; immunoglobulin.

Antigen A substance that triggers an immune response when it is introduced into the body.

Artificial immunity Immunity that requires some deliberate action, such as a vaccination, to achieve exposure to the potentially harmful antigen.

B-lymphocytes A cell of the immune system that develops into a plasma cell and produces antibodies; B cell.

Immunoglobulins Substances produced by the body that inactivate or destroy another substance that is introduced into the body; antibodies.

Lymph Fluid, which is derived from interstitial fluid, and found in the lymphatic vessels.

Lymph node A small, bean-shaped aggregate of lymphoid tissue along a lymphatic vessel that filters the lymph before it is returned to the blood circulation.

Natural immunity Immunity acquired through normal processes of daily living.

Passive immunity Immunity that results when an individual receives the immune agents from some source other than his or her own body.

Primary response The initial reaction of the immune system to a specific antigen.

Resistance Body's ability to counteract the effects of pathogens and other harmful agents.

Secondary response Rapid and intense reaction to antigens on second and subsequent exposures due to memory cells.

Susceptibility Lack of resistance to disease.

T-lymphocytes Cells of the immune system that differentiate in the thymus gland and are responsible for cell mediated immunity; T cells.

☞ **Chapter Objectives**

Upon completion of this chapter the student should be able to:

1. State three functions of the lymphatic system.
2. List the components of the lymphatic system.
3. Describe the origin and fate of lymph.
4. Name the two large lymphatic ducts and identify the region of the body each one drains.
5. State three actions that provide pressure gradients for fluid flow in lymphatic vessels.
6. Describe the structure, function, and distribution of lymph nodes.
7. Identify three groups of tonsils and describe their location and function.
8. Describe the location and structure of the spleen and discuss its functions.
9. Locate the thymus and describe its role in the production of lymphocytes.
10. Define the terms pathogen, resistance, and susceptibility.
11. List four nonspecific mechanisms that provide resistance to disease.
12. List three types of barriers against microbial invasion.
13. Describe the actions of complement and interferon.
14. Name the two principal phagocytic cells that function in defense against disease.
15. List four signs and symptoms of inflammation and briefly describe the process of inflammation.
16. Describe three responses present in systemic inflammation that are not present in localized inflammation.
17. State the two characteristics of specific defense mechanisms and identify the two principal cells involved in specific resistance.
18. Define the term antigen.
19. Describe the development of T-lymphocytes and B-lymphocytes and state the quantity of each type.
20. List four subgroups of T-cells and briefly describe the mechanism of cell mediated immunity.
21. List two subgroups of B-cells and briefly describe the mechanism of antibody mediated immunity.
22. Distinguish between the primary response and the secondary response to a pathogen.

23. Define the term immunoglobulin.
24. List five classes of immunoglobulins and state the role each has in immunity.
25. Distinguish between active and passive immunity.
26. Distinguish between natural and artificial immunity.
27. Give examples of active natural immunity, active artificial immunity, passive natural immunity, and passive artificial immunity.

☞ **Chapter Outline/Summary**

Functions of the Lymphatic System
(Objective 1)
- The lymphatic system returns excess interstitial fluid to the blood.
- The lymphatic system absorbs fats and fat soluble vitamins from the digestive system.
- The lymphatic system provides defense against invading microorganisms and disease.

Components of the Lymphatic System
(Objectives 2-9)
Lymph (Objective 3)
- As soon as interstitial fluid enters lymphatic vessels, it is called lymph.
- Lymphatic vessels return lymph to the blood plasma.

Lymphatic vessels (Objectives 4 and 5)
- Lymphatic vessels carry fluid away from the tissues and return it to the venous system.
- The right lymphatic duct drains lymph from the upper right quadrant of the body. The thoracic duct, which begins in the abdomen as the cisterna chyli, drains the other 3/4 of the body.
- Lymphatic vessels have thin walls and have valves to prevent backflow of blood.
- Pressure gradients that move fluid through the lymphatic vessels come from skeletal muscle action, respiratory movements, and contraction of smooth muscle in vessel walls.

Lymphatic organs (Objectives 6-9)
- Lymph nodes (Objective 6)
  ○ Three areas in which lymph nodes tend to cluster are the inguinal nodes in the groin, axillary nodes in the armpit, and cervical nodes in the neck. There are no lymph nodes associated with the central nervous system.
  ○ Lymph nodes consist of dense masses of lymphocytes that are separated by spaces called lymph sinuses. Lymph enters a node through afferent vessels, filters through the sinuses, and leaves through an efferent vessel.
  ○ Lymph nodes filter and cleanse the lymph before it enters the blood.
- Tonsils (Objective 7)
  ○ Tonsils are clusters of lymphatic tissue in the region of the nose, mouth, and throat.
  ○ Tonsils provide protection against pathogens that may enter through the nose and mouth.
  ○ The pharyngeal tonsils, also called adenoids, are near the opening of the nasal cavity into the pharynx; palatine tonsils are near the opening of the oral cavity into the pharynx; and the lingual tonsils are at the base of the tongue, also near the opening of the oral cavity into the pharynx.
- Spleen (Objective 8)
  ○ The spleen is located in the upper right quadrant of the abdomen, posterior to the stomach.
  ○ The spleen is much like a lymph node, but much larger. It contains masses of lymphocytes and macrophages that are supported by a fibrous framework.
  ○ The spleen filters blood in much the same way the lymph nodes filter lymph. The spleen also is a reservoir for blood.
- Thymus (Objective 9)
  ○ The thymus is located anterior to the ascending aorta and posterior to the sternum.
  ○ The principal function of the thymus is the processing and maturation of T-lymphocytes. It also produces thymosin that stimulates the maturation of lymphocytes in other organs.

Resistance to Disease (Objectives 10-26)
- Disease producing organisms are called pathogens. The ability to counteract pathogens is resistance and a lack of resistance is susceptibility. (Objective 10)

Nonspecific defense mechanisms (Objectives 11-15)
- Barriers (Objective 12)
  ○ Barriers are factors that deter microbial invasion.
  ○ Barriers may be of a mechanical nature (unbroken skin), fluid (tears), or chemical (lysozymes).
- Chemical action (Objective 13)
  ○ If microorganisms succeed in passing through the barriers internal defenses, such as chemical action, respond.
  ○ Complement is a chemical defense that promotes phagocytosis and inflammation.

o Interferon has particular significance because it offers protection against viruses. It is produced by virus infected cells to provide protection for the neighboring cells.

- Phagocytosis (Objective 14)
  o Phagocytosis is the ingestion and destruction of solid particles by certain cells, particularly neutrophils and macrophages.
  o Neutrophils are small cells that are the first to migrate to an infected area.
  o Macrophages are monocytes that leave the blood and enter the tissue spaces. They phagocytize cellular debris. They also perform a cleansing action on the lymph and blood.
- Inflammation (Objectives 15 and 16)
  o Inflammation is characterized by redness, warmth, swelling, and pain.
  o Inflammation includes a series of events, outlines in Table 14-1, that occur in response to tissue damage. The overall purpose of inflammation is to destroy bacteria, cleanse the area of debris, and promote healing.
  o Systemic inflammation is characterized by leukocytosis, fever, and a dangerous decrease in blood pressure.

Specific defense mechanisms (Objectives 17-24)

- Specificity and memory are two features of specific defense mechanisms. The principal cells involved are lymphocytes and macrophages.
- Recognition of self versus nonself (Objective 18)
  o Proteins and other large molecules that are recognized as belonging to an individual's own body are interpreted as "self". Others are interpreted as "nonself".
  o Antigens are molecules that trigger an immune response. Usually they are foreign proteins that enter the body and are interpreted as nonself.
- Development of lymphocytes (Objective 19)
  o During fetal development, the bone marrow releases immature lymphocytes into the blood.
  o Some of the immature lymphocytes go to the thymus gland where they differentiate to become T-lymphocytes (T-cells). About 70% of the circulating lymphocytes are T-cells.
  o Lymphocytes that differentiate in some region other than the thymus are B-lymphocytes (B-cells). These account for about 30% of the circulating lymphocytes.
- Cell mediated immunity (Objective 20)
  o Cell mediated immunity is the result of T-cell action.
  o When an antigen enters the body, it is phagocytized by a macrophage, which presents the antigen to an appropriate T-cell. The T-cell responds by producing clones of T-cells.
  o Killer T-cells directly destroy the cells with the offending antigen; helper T-cells secrete substances that promote the immune response; suppressor T-cells inhibit the immune response and help regulate it; memory T-cells stimulate a faster and more intense response if the same antigen enters the body again.
- Antibody mediated immunity (Objectives 21-24)
  o B-cells are responsible for antibody mediated immunity.
  o When an antigen enters the body, a macrophage phagocytizes it and presents to the appropriate B-cell. The B-cell responds by forming clones of plasma cells and memory B-cells.
  o Plasma cells produce large quantities of antibodies that inactivate the invading antigens. This initial action is a primary response.
  o Memory B-cells launch a rapid and intense response against the antigen if it enters the body again. This is called a secondary response.
  o Antibodies belong to a class of proteins called globulins. Since they are involved in the immune response, they are called immunoglobulins.
  o IgA, IgG, IgM, IgE, and IgD are five classes of immunoglobulins. Each class has a specific role in immunity (Table 14-2).

Acquired Immunity (Objectives 25-27)

- Active immunity occurs when the individual's own body produced memory cells; passive immunity results when the immune agents are transferred into an individual. Natural immunity is acquired through normal activities; artificial requires some deliberate action.
- Active natural immunity results when a person is exposed to a harmful antigen, contracts the disease, and recovers.

- Active artificial immunity develops when a prepared antigen is deliberately introduced into the body (vaccination) and stimulates the immune system.
- Passive natural immunity results when antibodies are transferred from mother to child through the placenta or milk.
- Passive artificial immunity results when antibodies are injected into an individual.

☞ **Answers to Review Questions**

1. The lymphatic system returns excess interstitial fluid to the blood, absorbs fats and fat-soluble vitamins, and provides defense against disease.
2. Lymph, vessels that transport the lymph, and lymphatic organs that contain lymphoid tissue.
3. Lymph is the fluid in the lymphatic vessels. It is picked up from the interstitial fluid and returned to the blood plasma.
4. The thoracic duct drains the entire body except the upper right quadrant, which is drained by the right lymphatic duct.
5. Pressure gradients to move the lymph come from skeletal muscle action, respiratory movements, and contraction of the smooth muscle in the vessel walls.
6. Lymph enters lymph nodes through afferent lymphatic vessel, filters through the lymph sinuses, and leaves the node through an efferent vessel. This pathway filters and cleanses the lymph before it returns to the blood.
7. Tonsils are located around the entrances into the pharynx or throat. Pharyngeal tonsils are near the opening of the nasal cavity into the pharynx; palatine tonsils are near the opening of the oral cavity into the pharynx; and the lingual tonsils are on the posterior surface of the tongue near the opening into the pharynx. The tonsils provide protection against pathogens that may enter through the nose and mouth.
8. The abundance of lymphocytes in the white pulp characterizes the spleen as a lymphatic organ. The spleen filters and cleanses blood, removes old and damaged RBCs, produces lymphocytes, and serves as a reservoir for blood.
9. The thymus is most active during infancy and childhood then begins to decrease in size after puberty. It functions in the processing and maturation of T-lymphocytes. It also produces the hormone thymosin.

10. Resistance is the body's ability to counteract the effects of pathogens and other harmful agents.
11. Nonspecific defense mechanisms are directed against all pathogens and foreign substances regardless of their nature. Examples include mechanical and chemical barriers, phagocytosis, and inflammation .
12. Intact, or unbroken, skin and mucous membranes form effective mechanical barriers against the entry of pathogens.
13. Interferon offers protection against viruses. It is produced by virus-infected cells to provide protection for the neighboring cells.
14. Neutrophils and macrophages are the primary phagocytic cells in the body.
15. Redness (rubor), warmth (calor), pain (dolor), and swelling (tumor) are the four characteristics of localized inflammation. In systemic inflammation, leukocytosis, increased body temperature or fever, and increased capillary permeability and vasodilation are also present.
16. Memory.
17. An antigen is a molecule that is interpreted as nonself and triggers an immune response.
18. About 70% of the circulating lymphocytes are T-cells. These differentiate in the thymus.
19. Killer T-cells, Helper T-cells, Memory T-cells, and Suppressor T-cells
20. Plasma cells produce antibodies in humoral, or antibody mediated, immunity.
21. The secondary response is more rapid because memory cells are present that recognize the antigen and initiate an immediate response.
22. Ig = immunoglobulin.
23. IgG is most abundant in blood plasma.
24. In passive immunity, the antibodies are produced in another person or animal and are transferred to the person who was not previously immune.
25. Active artificial immunity.

☞ **Answers to Learning Exercises**

Functions of the Lymphatic System (Objective 1)
1. Return excess interstitial fluid to the blood
   Absorption of fats and fat soluble vitamins
   Defense against invading microorganisms
Components of the Lymphatic System (Objectives 2-9)
1. Lymph
   Lymphatic vessels
   Lymphatic organs

2. Interstitial fluid
   Blood plasma
   Lymph
   Interstitial fluid
   Lymph capillaries
   Lymphatic ducts
   Right lymphatic duct
   Thoracic duct
   Cisterna chyli
   Right lymphatic duct
   Thoracic duct
   Valves
   Skeletal muscle action
   Respiratory movements
   Contraction of muscle in vessel walls
3. Lymphocytes
   Lymph nodes
   Tonsils
   Spleen
   Thymus
   Lymph nodes
   Spleen
   Tonsils (pharyngeal)
   Thymus
   Spleen
   Tonsils (pharyngeal)
   Afferent lymphatic vessel
   Thymosin
   Spleen

Resistance to Disease (Objectives 10-27)

1. Pathogens
   Resistance
   Susceptibility
   Nonspecific mechanisms
   Specific mechanisms
   Immunity
2. Leukocytosis
   Fever
   Decrease in blood pressure
3. A (unbroken skin and mucous membranes)
   C (activates phagocytosis & inflammation)
   C (blocks replication of viruses)
   A (cilia action in respiratory tract)
   C (interferon)
   D (accompanied by swelling, heat, redness)
   B (ingestion & digestion of solid particles)
   C (action of complement)
   B (macrophages and neutrophils)
   D (aimed at localizing damage)
4. Specificity
   Memory
   Lymphocytes
   Macrophages
   Antigens
   T-lymphocytes (T-cells)

B-lymphocytes (B-cells)
T-cells
B-cells
B-cells
T-cells
Plasma cells
Immunoglobulins
Primary response
Killer T-cells
Helper T-cells
Suppressor T-cells
Memory T-cells
Plasma cells
Memory B-cells

5. IgG (most numerous antibody)
   IgE (responsible for allergies)
   IgA (found in breast milk, saliva)
   IgM (responsible for ABO reactions)
   IgG (major antibody in immune response)
   IgE (binds to mast cells)
   IgG (crosses placenta)
   IgM (causes agglutination of antigens)
6. D (antiserum is injected...)
   A (a person contracts a disease...)
   C (antibodies are transferred...)
   B (antigens deliberately introduced...)
   A (memory cells ...)
   D (antibodies injected...)
   C (IgA antibodies in mother's milk)
   B (mumps, diphtheria...vaccines)

☞ **Answers to Chapter Self-Quiz**

1. d
2. c
3. d
4. Thymus
5. A (intact skin)
   B (interferon and complement)
   A (lysozymes in tears)
   D (swelling and pain)
   C (neutrophils and macrophages)
   A (fluid flow)
6. The primary cells involved in specific resistance are **lymphocytes** and **macrophages**.
7. TB (memory cells)
   T   (cell mediated immunity)
   B   (plasma cells)
   B   (humoral immunity)
   B   (antibodies)
   T   (clones of helper and killer cells)
   TB (macrophage presents antigen)
   B   (immunoglobulins)
8. b

☞ Answers to Terminology Exercises

Tonsillectomy
Splenoplexy
Lymphostasis
Immunology
Thymitis

Tumor of the thymus
Enlarged spleen
Study of lymph vessels
Inflammation of lymph glands
Decreased number of lymphocytes

E (poison substance)
D (surgical removal of the thymus)
A (incision into a lymph gland)
C (condition of spleen congested with blood)
F (formation of lymph)

☞ Answers to Fun and Games

1. Tonsils
2. Thymus
3. Lymphocyte
4. Immunoglobulin
5. Interleukin (from clinical terms)
6. Macrophage
7. Anaphylaxis
8. Neutrophil
9. Antigen
10. Lysozyme
11. Plasma cell
12. Intact skin
13. Interferon
14. Vaccination
15. Rubor
16. Phagocytosis
17. Lymphadenitis
18. Active
19. Pyrogens
20. Susceptibility

☞ Quiz/Test Questions

Note: There are fifty multiple-choice questions for this chapter in the computerized test bank.

Name the following:

1. Smallest lymphatic vessels.
   Answer: lymph capillaries.
2. Vessels that drains lymph from the upper right quadrant of the body.
   Answer: right lymphatic duct.
3. Structures that filter lymph.
   Answer: lymph nodes.

4. Clusters of lymphatic tissue under the mucous membrane of the nose, mouth, and throat.
   Answer: tonsils.
5. Largest lymphatic organ in the body; also filters blood.
   Answer: spleen.
6. Lack of resistance to disease.
   Answer: susceptibility.
7. Effective mechanical barrier against entry of pathogens.
   Answer: unbroken skin and mucous membranes.
8. Produced by virus-infected cells to provide protection for surrounding cells.
   Answer: interferon.
9. Substance that induces fever in systemic inflammation.
   Answer: pyrogen.
10. Tissue where T-lymphocytes differentiate and mature.
    Answer: thymus.
11. Four clones of T-lymphocytes.
    Answer: killer T-cells, suppressor T-cells, helper T-cells, memory T-cells.
12. Cell that ingests and processes antigens in both cell-mediated and humor immunity.
    Answer: macrophage.
13. Major antibody in primary and secondary immune responses.
    Answer: IgG.
14. Type of immunity obtained when an infant receives IgA antibodies from mother's milk.
    Answer: passive natural immunity.
15. B-cell clone that produces antibodies.
    Answer: plasma cell.

True/False Questions:

1. There are generally more afferent vessels associated with a lymph node than efferent vessels.
   Answer: True; this slows the flow of lymph to allow time for cleansing.
2. The spleen filters lymph in addition to serving as a reservoir for blood.
   Answer: False; it filters blood, lymph is filtered by lymph nodes.
3. A molecule, generally a protein, that triggers an immune response is an antigen.
   Answer: True.

Questions for this chapter continue on page 139.

15 Respiratory System

☞ Key Terms/Concepts

Alveolus Small sac-shaped structure; most often used to denote the microscopic dilations of terminal bronchioles in the lungs where diffusion of gases occurs; air sacs in the lungs.

Bronchial tree The bronchi and all their branches that function as passageways for between the trachea and the alveoli.

Bronchopulmonary segment Portion of a lung surrounding a tertiary, or segmental, bronchus; lobule of the lung.

Carina Ridge of hyaline cartilage in the region where the trachea divides into the right and left bronchi.

Epiglottis Long, leaf-shaped, movable cartilage of the larynx that covers the opening of the larynx and prevents food from entering during swallowing.

Expiration Process of letting air out of the lungs during the breathing cycle; also called exhalation.

External respiration Exchange of gases between the lungs and the blood.

Inspiration Process of taking air into the lungs; also called inhalation.

Internal respiration Exchange of gases between the blood and tissue cells.

Oxyhemoglobin Compound that is formed when oxygen binds with hemoglobin; form in which most of the oxygen is transported in the blood.

Pleural cavity The small space between the parietal and visceral layers of the pleura.

Respiratory membrane Surfaces in the lungs where diffusion occurs and consists of the layers that the gases must pass through to get into or out of the alveoli.

Surfactant A substance produced certain cells in lung tissue that reduces surface tension between fluid molecules that line the respiratory membrane and helps keep the alveolus from collapsing.

Ventilation Movement of air into and out of the lungs; breathing.

☞ Chapter Objectives

Upon completion of this chapter the student should be able to:

1. Define five activities of the respiratory process.
2. Distinguish between the upper respiratory tract and the lower respiratory tract.
3. Label a diagram showing the features of the nasal cavities.
4. List three functions of the nasal cavities.
5. Name the three regions of the pharynx, state where each is located, and identify the features unique to each region.
6. Describe the location of the larynx and identify the 3 largest cartilages in its framework.
7. Describe the framework of the trachea and identify the type of tissue that forms the lining.
8. List in sequence the branches of the bronchial tree beginning with the bifurcation of the trachea and ending with the alveolar ducts.
9. Name the type of tissue that forms the alveoli.
10. Compare the right lung and left lung by shape and number of lobes.
11. Define parietal pleura, visceral pleura, and pleural cavity.
12. Name and define three pressures involved in pulmonary ventilation.
13. Describe the role of the diaphragm in inspiration and expiration.
14. List the sequence of events that result in inspiration and expiration.
15. Explain the importance of surfactant in maintaining inflated alveoli.
16. Define 4 respiratory volumes and 4 respiratory capacities and state their average normal values.
17. Describe 4 factors that may influence lung volumes and capacities.
18. State Dalton's law of partial pressures and Henry's gas law.
19. List the 6 layers of the respiratory membrane through which gases diffuse in external respiration.
20. Discuss 3 factors that affect the rate at which external respiration occurs.
21. Distinguish between external respiration and internal respiration.

22. Describe two ways in which oxygen is transported in the blood.
23. Discuss 3 mechanisms of carbon dioxide transport in the blood.
24. Name 2 regions in the brain that make up the respiratory center and 2 nerves that carry impulses from the center.
25. Describe the role of chemoreceptors, stretch receptors, higher brain centers, and temperature in regulating breathing.

☞ **Chapter Outline/Summary**

Functions and Overview of Respiration (Objective 1)
- The entire process of respiration includes ventilation, external respiration, transport of gases, internal respiration, and cellular respiration.

Ventilation (Objectives 2-17)

Conducting passages (Objectives 2-11)
- The upper respiratory tract includes the nose, pharynx, and larynx. The lower respiratory tract consists of the trachea, bronchial tree, and lungs.
- Nose and nasal cavity (Objectives 3 and 4))
  - The nasal cavity opens to the outside through the external nares and into the pharynx through the internal nares. It is separated from the oral cavity by the palate.
  - The frontal, maxillary, ethmoidal, and sphenoidal sinuses are air filled cavities that open into the nasal cavity.
  - Air is warmed, moistened, and filtered as it passes through the nasal cavity.
- Pharynx (Objective 5)
  - The region of the pharynx is divided into the nasopharynx (posterior to the nasal cavity), oropharynx (posterior to the oral cavity), and the laryngopharynx (posterior to the larynx).
  - Pharyngeal tonsils are located in the wall of the nasopharynx and the auditory tubes open into the nasopharynx.
  - The opening from the oral cavity into the oropharynx is called the fauces. The palatine and lingual tonsils are located in this region.
- Larynx (Objective 6)
  - The larynx is formed by nine cartilages that are connected to each other by muscles and ligaments.
  - The three largest cartilages of the larynx are the thyroid cartilage, cricoid cartilage, and the epiglottis.
  - There are two pair of folds in the larynx. The upper pair are the vestibular folds. The lower pair are the true vocal cords and the opening between the vocal cords is the glottis.
- Trachea (Objective 7)
  - The framework of the trachea is supported by 15-20 C-shaped pieces of hyaline cartilage.
  - The mucous membrane that lines the trachea has goblet cells and cilia. The goblet cells secrete mucus that traps particles that are inhaled. The cilia provide a cleansing action to remove the mucus with the particles.
- Bronchi (Objectives 8 and 9)
  - The trachea divided into the right and left primary bronchi.
  - The primary bronchi divided into secondary (lobar) bronchi, and these into tertiary (segmental) bronchi.
  - The branching pattern continues into smaller and smaller passageways until they terminate in tiny air sacs called alveoli.
  - The alveoli consist of simple squamous epithelium, which permits rapid diffusion of oxygen and carbon dioxide.
- Lungs (Objectives 10 and 11)
  - The right lung is shorter, broader, and has a greater volume than the left lung.
  - The right lung is divided into 3 lobes; the left lung has 2 lobes.
  - The left lung has an indentation, called the cardiac notch, for the apex of the heart.
  - The parietal pleura lines the wall of the thorax; the visceral pleura is firmly attached to the surface of the lung. The pleural cavity is the space between the two layers of pleura.

Mechanics of ventilation (Objectives 12-15)
- Pressures in pulmonary ventilation (Objective 12)
  - Air flows because of pressure differences between the atmosphere and the gases inside the lungs.
  - Atmospheric pressure is the pressure of the air outside the body.
  - Intraalveolar (intrapulmonary) pressure is the pressure inside the alveoli of the lungs.
  - Intrapleural pressure is the pressure within the pleural cavity, between the visceral and parietal pleurae.

- Inspiration (inhalation) (Objectives 13 and 14)
  - During inspiration, the diaphragm contracts and the thoracic cavity increases in volume.
  - An increase in thoracic volume decreases the pressure below atmospheric pressure so that air flows into the lungs.
- Expiration (exhalation) (Objectives 13-15)
  - During expiration, the relaxation of the diaphragm and elastic recoil of tissues decrease the thoracic volume.
  - The decrease in thoracic volume increases the intraalveolar pressure so that air flows out of the lungs.
  - Surfactant reduces the surface tension inside the alveoli so they do not adhere to each other and collapse.

Respiratory volumes and capacities (Objectives 16 and 17)

- A spirometer is used to measure the volume of air that moves into and out of the lungs.
- The four respiratory volumes measured by spirometry are the tidal volume, inspiratory reserve volume, expiratory reserve volume, and residual volume.
- A respiratory capacity is the sum of two or more volumes. Four respiratory capacities are the vital capacity, inspiratory capacity, functional residual capacity, and total lung capacity.
- Age, sex, body build, and physical conditioning have an influence on lung volumes and capacities.

Basic Gas Laws and Respiration (Objectives 18-21)

Properties of gases (Objective 18)

- Dalton's law of partial pressures states that the total pressure exerted by a mixture of gases is equal to the sum of the pressures exerted by each gas independently and the partial pressure exerted by each gas is proportional to its percentage in the total mixture.
- Henry's law states that when a mixture of gases is in contact with a liquid, each gas dissolves in the liquid in proportion to its own solubility and partial pressure.

External respiration (Objectives 19-21)

- The surfaces in the lungs where diffusion occurs are called the respiratory membranes.
- The layers of the respiratory membrane are the (1) thin layer of fluid that lines the alveolus, (2) simple squamous epithelium in the alveolar wall, (3) basement membrane of the epithelium, (4) small interstitial space, (5)

basement membrane of capillary epithelium, and (6) simple squamous epithelium of the capillary wall.

- The rate at which external respiration occurs varies with the surface area and thickness of the respiratory membrane, the solubility of the gas, and the difference in partial pressure of the gas on the two sides of the membrane.

Internal Respiration (Objective 21)

- Internal respiration is the exchange of gases between the tissue cells and the blood in the tissue capillaries.
- Oxygen diffuses from the blood into the tissue cells and carbon dioxide diffuses from the tissue cells into the blood during internal respiration.

Transport of Gases (Objectives 22 and 23)

Oxygen (Objective 22)

- Approximately 3% of the oxygen is transported as a dissolved gas in the plasma. The remaining 97% is carried by hemoglobin molecules as oxyhemoglobin.
- Loading occurs in the lungs when oxygen combines with hemoglobin. Unloading occurs in the tissues when hemoglobin releases oxygen.
- Loading takes place when oxygen levels are high and carbon dioxide levels are low. Unloading occurs when oxygen levels are low, carbon dioxide levels are high, temperature is increased, and pH is decreased.

Carbon dioxide (Objective 23)

- Approximately 7% of the carbon dioxide is transported as a gas dissolved in the plasma.
- Another 23% of the carbon dioxide combines with the protein portion of hemoglobin and is transported as carbaminohemoglobin.
- The remaining 30% of the carbon dioxide is transported as bicarbonate ions in the plasma.
- In the lungs, where carbon dioxide levels are relative low, reactions occur that release the carbon dioxide from its transport forms. The carbon dioxide diffuses into the alveoli and is exhaled.

Regulation of Respiration (Objectives 24 and 25)

Respiratory center (Objective 24)

- The respiratory center includes groups of neurons in the medulla oblongata and pons.
- The respiratory center contains both inspiratory and expiratory neurons.
- The inspiratory area sends impulses along the phrenic nerve to the diaphragm and along the intercostal nerves to the external intercostal muscles.

- When inspiratory impulses cease, the muscles relax, and expiration occurs. When more forceful expiration is necessary, the expiratory center sends impulses along the intercostal nerves to the internal intercostal muscles.

Factors that influence breathing (Objective 25)

- Central chemoreceptors in the medulla oblongata are sensitive to increases in carbon dioxide and hydrogen ion levels.
- Peripheral chemoreceptors in the aortic and carotid bodies detect decreases in oxygen levels but this is not a strong stimulus for breathing.
- Stretch receptors in the lungs initiate the Hering-Breuer reflex that prevents over-inflation of the lungs.
- Voluntary or involuntary impulses from the higher brain centers may override the respiratory center temporarily but after a limited time, the respiratory center resumes control.
- An increase in temperature increases the breathing rate.

☞ Answers to Review Questions

1. Ventilation, external respiration, transport of gases, internal respiration, cellular respiration.
2. Upper respiratory tract: nose, pharynx, larynx.
   Lower respiratory tract: trachea, bronchial tree, lungs.
3. External nares are the nostrils, or the openings from the outside into the nasal cavity.
   Internal nares are the openings from the nasal cavity into the pharynx.
   The hard palate is the anterior portion of the roof of the mouth, the portion that is supported by bone.
   The soft palate is the posterior portion of the roof of the mouth, the portion that has no bony support.
   The uvula is the terminal portion of the soft palate and functions to help direct food into the oropharynx.
   Nasal conchae are bony ridges that project medially into the nasal cavity from each lateral wall.
4. The nasal cavity warms, filters, and moistens the air.
5. Pharyngeal tonsils are in the nasopharynx; palatine and lingual tonsils are in the oropharynx.
6. Epiglottis, thyroid, and cricoid.
7. Hyaline cartilage.
8. Mucus from the goblet cells in the mucous membrane traps foreign particles. Cilia move the mucus with trapped particles upward and out of the tract.
9. Trachea → primary bronchi → secondary bronchi → tertiary bronchi → bronchioles → terminal bronchioles → respiratory bronchioles → alveolar ducts → alveoli.
10. They are composed of simple squamous epithelium which is very thin.
11. Right lung has 3 lobes; left lung has 2 lobes.
12. The pleural cavity is between the parietal pleura and visceral pleura. This is a potential space that contains a thin film of serous fluid for lubrication.
13. Atmospheric pressure is the greatest.
14. Inspiration is an active process because it involves contraction of the diaphragm. Expiration involves the relaxation of the diaphragm.
15. Surfactant.
16. See Figure 15-6 in the textbook.
17. Lung volumes usually decline after early adulthood; females generally have less volume than males; tall people have greater volume than short people; thin people have greater volume than obese people; physical conditioning increases lung capacities.
18. Dalton's law of partial pressures and Henry's law.
19. Thin layer of fluid that lines the alveolus
    Simple squamous epithelium in alveolar wall
    Basement membrane of the epithelium
    Small interstitial space
    Basement membrane of capillary epithelium
    Simple squamous epithelium of the capillary wall
20. The rate of external respiration is decreased.
21. Internal respiration is the exchange of gases between the tissue cells and the blood in the tissue capillaries.
22. An increase in carbon dioxide concentration decreases the ability of oxygen to combine with hemoglobin.
23. An increase in carbon dioxide provides more stimulus for breathing.
24. As bicarbonate ions.
25. In the pons and medulla oblongata.
26. Phrenic nerve sends impulses to the diaphragm.
    Intercostal nerves send impulses to the external intercostal muscles.

☞ **Answers to Learning Exercises**

Functions   and   Overview   of   Respiration (Objective 1)
1. Ventilation
    External respiration
    Transport
    Internal respiration
    Cellular respiration

Ventilation (Objectives 2-17)
1. F (ethmoidal sinus)
    E (frontal sinus)
    H (hard palate)
    G (nasal cavity)
    B (opening for auditory tube)
    I (oral cavity)
    D (pharynx)
    C (soft palate)
    A (sphenoidal sinus)
2. L (bronchi)
    U (larynx)
    L (lungs)
    U (nose)
    U (pharynx)
    L (trachea)
3. Warm the air
    Filter the air
    Moisten the air
    Nasopharynx
    Nasopharynx
    Oropharynx
    Laryngopharynx
    Thyroid
    Cricoid
    Epiglottis
    Glottis (rima glottis)
    Trachea
    Carina
    Pseudostratified ciliated columnar epith.
    Simple squamous epithelium
4. B (carina)
    E (cricoid cartilage)
    F (primary bronchus)
    C (secondary bronchus)
    D (thyroid cartilage)
    A (trachea)
5. 6 (alveolar ducts)
    7 (alveoli)
    2 (lobar bronchi)
    1 (primary bronchi)
    5 (respiratory bronchioles)
    3 (segmental bronchi)
    4 (terminal bronchioles)
6. L (cardiac notch)
    R (shorter and wider)
    B (rests on the diaphragm)

B (enclosed by the pleura)
L (two lobes)
B (divided into lobules)
R (has two fissures)
B (anchored at the root or hilum)

7. Intrapleural pressure
    Atmospheric pressure
    Intrapulmonary pressure
    Intrapleural pressure
    Diaphragm
    Internal intercostals
    Surfactant
    Spirometer
    Intrapulmonary pressure
    Atmospheric pressure
8. I (diaphragm contracts)
    E (intrapulmonary exceeds atmospheric)
    I (external intercostal muscles may contract)
    I (atmospheric greater than intrapulmonary)
    I (lung volume increases)
    E (diaphragm relaxes)
    E (internal intercostal muscles may contract)
    I (air flows into the lungs)
    E (elastic recoil decreases size of alveoli)
9. IC = Inspiratory capacity
    TV = Tidal volume
    FRC = Functional residual capacity
    VC = Vital capacity
    IRV = Inspiratory reserve volume
    TLC = Total lung capacity
    ERV = Expiratory reserve volume
    FRC = Functional residual capacity
    RV = Residual volume
    TLC = Total lung capacity
10. First row: 5700 ml TLC and 4500 ml VC
    Second row: 600 ml TV and 1000 ml RV
    Third row: 5800 ml TLC and 3300 ml IRV
    Fourth row: 1200 ml ERV and 1300 ml RV
11. Young adults > (greater than)
    Females < (less than)
    Short people < (less than)
    Normal weight people > (greater than)
    Healthy people > (greater than)
    Good physical condition > (greater than)

Basic Gas Laws and Respiration (Objectives 18-21)
1. 21% x 750 mm Hg = 157.5 mm Hg
2. Solubility of each gas and partial pressures
3. External respiration is the exchange of gases between the **alveoli in the lungs and the blood in the capillaries.** Internal respiration is the exchange of gases between the **blood in the lungs and the tissue cells.**
4. a. Fluid that lines the alveolus
    b. Simple squamous epith. of alveolar wall
    c. Basement membrane of epithelium

d. Interstitial space

e. Basement membrane of capillary epith.

f. Simple squamous epith. of capillary wall

5. D

D

I

D

Transport of Gases (Objectives 22 and 23)

1. Dissolved in plasma and bound to hemoglobin as oxyhemoglobin

2. Dissolved in plasma

   Bound to Hb as carbaminohemoglobin

   As bicarbonate ions in plasma

3. Oxyhemoglobin

   Bicarbonate ions

   Carbaminohemoglobin

   Carbonic acid

   Catalase

4. Should have checks (✓) before the following:

   Increased partial pressure of carbon dioxide

   Increased hydrogen ion concentration

   Increased temperature

   Increased cellular metabolism

5. E (oxygen diffuses into blood)

   I (oxygen diffuses out of blood)

   I (carbon dioxide diffuses into blood)

   E (carbon dioxide diffuses out of blood)

   E (occurs in alveolus)

   I (occurs in body tissues)

   I (bicarbonate ion is formed)

   E (bicarbonate ions release carbon dioxide)

   E (oxyhemoglobin is formed)

   I (oxyhemoglobin dissociates)

Regulation of Respiration
(Objectives 24 and 25)

1. The respiratory center includes neurons in the **pons** and **medulla oblongata**.

2. The inspiratory areas of the respiratory center send impulses along the **phrenic** nerve to the diaphragm and along the **intercostal** nerves to the external intercostal muscles.

3. F (Chemoreceptors in the medulla...)

   T (Increases in blood...)

   T (Increases in hydrogen ion...)

   T (Chemoreceptors in the medulla...)

   F (A decrease in oxygen...)

   T (Decreased oxygen levels...)

   T (Peripheral chemoreceptors...)

   T (The Hering-Breuer reflex...)

   T (The Hering-Breuer reflex is...)

   F (Higher brain centers...)

   T (Anxiety decreases...)

   T (Chronic pain stimulates...)

   F (Decreasing body temperature...)

   T (The primary stimulus...)

☞ Answers to Chapter Self-Quiz

1. b

2. d

3. F

   F

   T

   T

4. b

5. d

6. a

7. b

8. d

9. d

10. c

☞ Answers to Terminology Exercises

Apnea

Cricoid

Rhinoplasty

Bronchiectasis

Hyperpnea

Spitting blood

Resembling a shield

Inflammation of the nose

Difficult breathing

Condition of dust in the lungs

E (discharge from the nose)

A (presence of tiny cavities)

C (process of voice production)

B (condition caused by inhaling coal dust)

D (surgical removal of a lung)

☞ Answers to Fun and Games

1. Pharynx (+)

2. Pleura (+)

   Letters: pleurapharynx

3. Larynx (−)

   Letters: peuphar

4. Surfactant (+)

   Letters: peupharsurfactant

5. Os (+)

   Letters: surfactant peupharos

6. Apnea (−)

   Letters: surfacttpuhros

7. Nares (+)

   Letters: surfacttpuhrosnares

8. Fauces (−)

   Letters: srttpuhrosnar

9. A (−)

   Letters: srttpuhrosnr

10. Carina (+)

    Letters: srttpuhrosnrcarina

11. Alveolus (+)
    Letters: srttpuhrosnrcarinaalveolus
12. Alp (−)
    Letters: srttuhrosnrcarinaveolus
13. Trachea (−)
    Letters: sturosnrrinvolus
14. Atelectasis (+)
    Letters: sturosnrrinvolusatelectasis
15. Diaphragm (+)
    Letters: sturosnrrinvolusatelectasisdiaphragm
16. Rang (−)
    Letters: sturosnrivolustelectasisdiaphram
17. Pertussis (−)
    ronvolutlectasisdiaramh
18. Tidal volume (−)
    Letters: ronctsisarah
19. Anthracosis (−)
    Letter: r
    FINAL LETTER: R for Respiration

☞ **Quiz/Test Questions**

Note: There are fifty multiple-choice questions for this chapter in the computerized test bank.

Name the following:

1. First, or uppermost, portion of the lower respiratory tract.
   Answer: trachea.
2. Bony ridges that project medially from the lateral wall of the nasal cavity.
   Answer: nasal conchae or turbinates.
3. Openings from the nasal cavity into the nasopharynx.
   Answer: internal nares or choanae.
4. Laryngeal cartilage that covers the opening of the larynx during swallowing.
   Answer: epiglottis.
5. Type of tissue that forms the lining of the trachea.
   Answer: pseudostratified ciliated columnar epithelium.
6. Air sacs where diffusion of gases occurs.
   Answer: alveoli.
7. Number of secondary, or lobar, bronchi to the right lung.
   Answer: three.
8. Pressure that always has the least value of the three involved in ventilation.
   Answer: intrapleural pressure.
9. Term given to the air that is normally inhaled and exhaled in quiet breathing.
   Answer: tidal volume.

10. Type of epithelium in the respiratory membrane.
    Answer: simple squamous epithelium.
11. Form in which 97 percent of the oxygen is transported.
    Answer: oxyhemoglobin.
12. Form in which most of the carbon dioxide is transported.
    Answer: bicarbonate ions.
13. Nerve that carries impulses to the diaphragm.
    Answer: phrenic nerve.
14. Two regions of the CNS that contain respiratory neurons.
    Answer: pons and medulla oblongata.
15. Factor that provides the strongest stimulus for breathing.
    Answer: increase in carbon dioxide concentration or increase in hydrogen ion concentration.

True/False Questions:

1. Air enters the lungs when intrapulmonary pressure is greater than atmospheric pressure.
   Answer: False; air enters when atmospheric pressure is greater than intrapulmonary pressure.
2. The support for the tracheal wall comes from rings of fibrocartilage.
   Answer: False; it is hyaline cartilage.
3. The right lung is shorter and wider than the left lung and has three lobes.
   Answer: True.
4. The removal of the pharyngeal tonsils is called a tonsillectomy.
   Answer: False; pharyngeal tonsils are the adenoids.
5. The parietal pleura is closely adherent to the surface of the lungs.
   Answer: False; the visceral pleura is closely adherent to lung surface.
6. Active transport accounts for the exchange of air between the lungs and the blood.
   Answer: False; the exchange is by diffusion.
7. When the diaphragm contracts, the volume of the lungs increases, and the intraalveolar pressure decreases.
   Answer: True.
8. Surfactant makes the inner surfaces of the alveoli tend to stick together.
   Answer: False; it reduces the surface tension so they are easier to inflate.

Questions for this chapter continue on page 139.

16 Digestive System

☞ Key Terms/Concepts

Absorption The passage of digestive end products from the gastrointestinal tract into the blood or lymph.

Bile Yellowish-green fluid that is produced by the liver, stored in the gallbladder, and functions to emulsify fats.

Chylomicrons Small fat droplets that are covered with a protein coat in the epithelial cells of the mucosa of the small intestine.

Chyme The semifluid mixture of food and gastric juice that leaves the stomach through the pyloric sphincter.

Defecation The removal of indigestible wastes, or feces, through the anus.

Deglutition The process of swallowing.

Feces Material discharged from the rectum consisting of bacteria, indigestible food residue, and secretions.

Gastrointestinal tract Long, continuous tube that extends from the mouth to the anus; the digestive tract or alimentary canal.

Hepatocytes Liver cells.

Ingestion The process of taking in food.

Mastication The process of chewing.

Mesentery Extensions of peritoneum that are associated with the intestine.

Micelles Tiny droplets of monoglycerides and free fatty acids that are coated with bile salts before absorption.

Peristalsis Rhythmic contractions of the intestines that move food along the digestive tract.

Plicae circulares Circular folds in the mucosa and submucosa of the small intestine .

Rugae Longitudinal folds in the mucosa of the stomach.

Teniae coli Bands of longitudinal muscle fibers in the large intestine.

☞ Chapter Objectives

Upon completion of this chapter the student should be able to:

1. List the components of the digestive tract and the accessory organs.
2. List six functions of the digestive system.
3. Describe the general structure of the four layers, or tunics, in the wall of the digestive tract.
4. List and describe the boundaries of the oral cavity.
5. Describe the structure and functions of the tongue.
6. Distinguish between the primary and secondary teeth.
7. Show the relationship between tooth shape and function.
8. Identify the features of a "typical" tooth.
9. Name and describe the location of the three major types of salivary glands.
10. List the components and describe the functions of saliva.
11. Identify and describe the features of the pharynx.
12. Describe the location and features of the esophagus.
13. Identify the features of the stomach.
14. Describe how the structure of the stomach wall differs from the generalized structure of the digestive tract.
15. Name the types of cells found in gastric glands and tell what is secreted by each type.
16. Describe the events in each of the three phases in the regulation of gastric secretion.
17. Discuss two factors that affect the rate at which the stomach contents enter the small intestine.
18. Describe three features of the small intestine that increase the surface area for absorption.
19. Distinguish between the three regions of the small intestine with respect to location, length, and features.
20. Name 5 digestive enzymes and 2 hormones that are produced in the small intestine and indicate the function of each one.
21. Describe how chyme affects intestinal secretions.
22. Describe how the structure of the wall of the large intestine differs from the generalized structure of the digestive tract.
23. List the regions of the large intestine and state the location of each region.
24. List two functions of the large intestine.
25. Identify the external features of the liver.

26. Draw and label a diagrammatic representation of a liver lobule.
27. Distinguish between the two sources of blood for the liver and trace the flow of blood through the liver.
28. List 10 functions of the liver.
29. Describe the role of bile in digestion and as an excretory agent.
30. Describe the location and function of the gallbladder.
31. Describe the location of the pancreas.
32. Distinguish between the exocrine and endocrine portions of the pancreas.
33. List four pancreatic enzymes and explain the function of each one.
34. Name two hormones that regulate pancreatic secretions and distinguish between the functions of each one.
35. Summarize carbohydrate digestion by writing an equation that shows the intermediate and final products and the enzymes that facilitate the digestive process.
36. Summarize protein digestion by writing an equation that shows the intermediate and final products and the enzymes that facilitate the digestive process.
37. Summarize lipid digestion by writing an equation that shows the intermediate and final products and the factors that facilitate the digestive process.
38. Compare the absorption of simple sugars and amino acids with that of lipid related molecules.

☞ Chapter Outline/Summary

Introduction (Objective 1)
- The digestive tract includes the mouth, pharynx, esophagus, stomach, small intestine, and large intestine.
- The accessory organs of the digestive system are the salivary glands, liver, gallbladder, and pancreas.

Functions of the Digestive System (Objective 2)
- The digestive system prepares nutrients for utilization by the cells of the body.
- Activities of the digestive system include ingestion, mechanical digestion, chemical digestion, mixing and propelling movements, absorption, and elimination of waste products.

General Structure of the Digestive Tract (Objective 3)
- The basic structure of the wall of the digestive tube is the same throughout the entire length, although there are variations in each region.

- The wall of the digestive tract consists of a mucosa, submucosa, muscular layer, and an outer adventitia (above the diaphragm) or serosa (below the diaphragm).
- Nerve plexuses are located in the submucosa and in the muscular layer.

Regions of the Digestive Tract (Objectives 4-24)

Mouth (Objectives 4-10)
- Lips, Palate, and tongue (Objectives 4 and 5)
  ○ The lips and cheeks are muscles covered with epithelium and lined with mucous membrane.
  ○ The palate is the roof of the mouth. The anterior portion is supported by bone; the posterior portion is muscle and connective tissue.
  ○ The tongue is composed of skeletal muscle. The dorsal surface is covered with papillae, some of which contain taste buds. The tongue manipulates food in the mouth, contains sensory receptors for taste, and is used in speech.
- Teeth (Objectives 6-8)
  ○ The primary teeth are the deciduous teeth that fall out and are replaced by the secondary or permanent teeth.
  ○ There are 20 teeth in the complete primary set and 32 teeth in the complete secondary set.
  ○ The incisors have sharp edges for biting; cuspids have points for grasping and tearing; bicuspids and molars have flat surfaces for grinding.
  ○ Each tooth has a crown, a neck, and a root. Enamel covers the crown.
- Salivary glands (Objectives 9 and 10)
  ○ The parotid, submandibular, and sublingual glands secrete saliva, which contains the enzyme amylase.
  ○ The parotid glands are anterior and inferior to the ear; the submandibular glands are along the medial surface of the mandible; and the sublingual glands are under the tongue.
  ○ Saliva contains water, mucus, and amylase.
  ○ Saliva has a cleansing action, moistens food, dissolves substances for taste, and begins digestion of carbohydrates.

Pharynx (Objective 11)
- The pharynx is a passageway that transports food to the esophagus.
- The pharynx is divided into the nasopharynx, oropharynx, and laryngopharynx.

Esophagus (Objective 12)
- The esophagus is posterior to the trachea and anterior to the vertebral column.
- The lower esophageal sphincter, also called the cardiac sphincter, controls the passage of food into the stomach.

Stomach (Objectives 13-17)
- Structure (Objectives 13 and 14)
  - The stomach is divided into a fundus, cardiac region, body, and pyloric region and has a greater curvature and a lesser curvature.
  - The mucosal lining has folds called rugae and there are three layers of smooth muscle in the wall.
- Gastric secretions (Objective 15)
  - Mucous cells secrete mucus; parietal cells secrete hydrochloric acid and intrinsic factor; chief cells secrete pepsinogen; and endocrine cells secrete gastrin.
  - The semifluid mixture of food and gastric juice that leaves the stomach is called chyme.
- Regulation of gastric secretions (Objective 16)
  - The regulation of gastric secretions is divided into cephalic, gastric, and intestinal phases.
  - Thoughts and smells of food start the cephalic phase; the presence of food in the stomach initiates the gastric phase; and the presence of acid chyme in the small intestine starts the intestinal phase.
- Stomach emptying (Objective 17)
  - Relaxation of the pyloric sphincter allows chyme to pass from the stomach into the small intestine.
  - The rate at which stomach emptying occurs depends on the nature of the chyme and the receptivity of the small intestine.

Small intestine (Objectives 18-21)
- Structure (Objectives 18 and 19)
  - The absorptive surface area of the small intestine is increased by plicae circulares, villi, and microvilli.
  - Each villus contains a blood capillary network and a lymph capillary called a lacteal.
  - The small intestine is divided into the duodenum, jejunum, and ileum.
  - The duodenum has mucous glands in the submucosa; the jejunum has numerous, long villi; and the ileum has a large number of goblet cells.

- Secretions of the small intestine (Objectives 20 and 21)
  - Cells in the small intestine produce peptidase, which act on proteins; maltase, sucrase, and lactase, which act on disaccharides; and lipase, which acts on neutral fats.
  - The small intestine produces two hormones, secretin and cholecystokinin. Secretin stimulates the pancreas and cholecystokinin stimulates the gallbladder and digestive enzymes from the pancreas.
  - The presence of chyme in the duodenum stimulates intestinal secretions.

Large intestine (Objectives 22-24)
- Characteristic features (Objective 22)
  - The mucosa does not have villi, but has a large number of goblet cells.
  - The longitudinal muscle layer is limited to three bands called Teniae coli. Haustra and epiploic appendages are also characteristic.
- Regions (Objective 23)
  - The large intestine consists of the cecum, colon, rectum, and anal canal.
  - The colon is divided into the ascending colon on the right side, transverse colon across the anterior abdomen, descending colon on the left, and sigmoid colon across the pelvic brim.
- Functions (Objective 24)
  - The functions of the large intestine include the absorption of water and electrolytes and the elimination of feces.

Accessory Organs of Digestion (Objectives 25-34)
Liver (Objectives 25-29)
- Structure of the liver (Objectives 25 and 26)
  - Externally, the liver is divided into right, left, caudate, and quadrate lobes by the falciform ligament, INC, gallbladder, ligamentum venosum, and ligamentum teres.
  - The porta of the liver is where the hepatic artery and hepatic portal vein enter the liver and the hepatic ducts exit.
  - The functional units of the liver are lobules with sinusoids that carry blood from the periphery to the central vein of the lobule.
- Blood supply to the liver (Objective 27)
  - Blood is brought to the liver by the hepatic portal vein and the hepatic artery. The blood from both vessels flows through the sinusoids into the central vein.

- o Central veins merge to form the hepatic veins, which drain into the inferior vena cava.
- Functions of the liver (Objective 28)
  - o The liver has numerous functions that include secretion, synthesis of bile salts, synthesis of plasma proteins, storage, detoxification, excretion, carbohydrate metabolism, lipid metabolism, protein metabolism, and filtering blood.
- Bile (Objective 29)
  - o The main components of bile are water, bile salts, bile pigments, and cholesterol.
  - o Bile salts act as emulsifying agents in the digestion and absorption of fats.
  - o The principal bile pigment is bilirubin, which is formed from the breakdown of hemoglobin.
  - o Cholesterol and bile pigments are excreted from the body in the bile.

Gallbladder (Objective 30)

- The gallbladder is attached to the visceral surface of the liver by the cystic duct, which joins the hepatic duct to form the common bile duct. The common bile duct empties into the duodenum.
- The gallbladder stores and concentrates the bile.

Pancreas (Objectives 31-34)

- The pancreas is retroperitoneal along the posterior body wall and extends from the duodenum to the spleen.
- Most of the pancreas is exocrine and composed of acinar cells, which produce digestive enzymes. The islets of Langerhans are endocrine and produce insulin and glucagon.
- Pancreatic enzymes include amylase, which acts on starch; trypsin, which acts on proteins; peptidase, which acts on peptides; and lipase, which acts on lipids.
- The hormone secretin stimulates the pancreas to secrete a bicarbonate rich fluid; cholecystokinin stimulates the production of pancreatic enzymes.

Chemical Digestion (Objectives 35-37)

Carbohydrates (Objective 35)

- Carbohydrates are first broken down into disaccharides by amylase.
- Disaccharides are broken down into monosaccharides by sucrase, maltase, and lactase.
- The end products of carbohydrate digestion are glucose, fructose, and galactose.

Proteins (Objective 36)

- Pepsin and trypsin break proteins into shorter chains called peptides.
- Peptidase breaks peptides into amino acids.
- The end products of protein digestion are amino acids.

Lipids (Objective 37)

- Fats are first emulsified by bile.
- Lipase acts on emulsified fats and breaks them down into monoglycerides and free fatty acids.
- Monoglycerides and free fatty acids are the end products of lipid digestion.

Absorption (Objective 38)

- Most nutrient absorption takes place in the jejunum.
- Water is absorbed by osmosis in all regions.
- Simple sugars and amino acids are absorbed into the blood capillaries in the villi of the small intestine, then transported to the liver in the hepatic portal vein.
- Fatty acids, monoglycerides, and fat soluble vitamins enter the lacteals in the villi of the small intestine and circulate in the lymph until the lymph enters the left subclavian vein.

☞ **Answers to Review Questions**

1. Mouth, pharynx, esophagus, stomach, small intestine, and large intestine.
2. Ingestion, mechanical digestion, chemical digestion, movements through the tube, and absorption.
3. The four layers of the digestive tract wall are the mucosa, submucosa, muscular layer, and serous layer. The submocosal, or Meissner's, plexus is in the submucosa. The myenteric, or Auerbach's, plexus is in the muscular layer.
4. The oral cavity is lined with stratified squamous epithelium and the major muscle in the cheek is the buccinator.
5. The hard and soft palates make up the roof of the mouth and separate the oral cavity from the nasal cavity. The hard palate is supported by bone, but the soft palate is entirely soft tissue.
6. Skeletal muscle makes up the substance of the tongue.
7. There are 20 teeth in the complete set of primary teeth. These are also called deciduous teeth.
8. Incisors, cuspids or canines, bicuspids or premolars, and molars.
9. See Figure 16-5 in the textbook.
10. The largest salivary gland is the parotid, which is located between the skin and masseter muscle, just anterior and inferior to the

ear. The parotid gland produces salivary amylase.

11. Nasopharynx posterior to the nasal cavity, oropharynx posterior to the oral cavity, and laryngopharynx or hypopharynx posterior to the larynx.

12. The fauces is the opening from the oral cavity into the oropharynx.

13. The lower esophageal sphincter is a circular band of muscle at the lower end of the esophagus where it opens into the stomach. It is also called the cardiac sphincter.

14. The **cardiac region** is the area around the opening for the esophagus; the **fundus** is the portion that balloons upward from the esophageal opening; the **body** is the main portion; and the **pyloric region** is where the stomach narrows just before entering the small intestine.

15. In most other regions of the digestive tract there are two muscular layers, an inner circular layer and an outer longitudinal layer. The stomach has three layers. An oblique layer is internal to the other two layers.

16. The exocrine secretions of the stomach are **mucus** from the goblet cells and other types of mucous cells, **hydrochloric acid** and **intrinsic factor** from the parietal cells, and **pepsinogen** from the chief cells. **Gastrin** is a hormone secreted by the endocrine cells.

17. The hormone, **gastrin**, and cranial nerve X, **vagus**, regulate gastric juice secretion.

18. The **pyloric sphincter** relaxes to allow chyme to leave the stomach.

19. Plicae circulares, villi, and microvilli increase the absorptive area of the small intestine.

20. a. Jejunum
    b. Duodenum
    c. Ileum
    d. Duodenum
    e. Duodenum
    f. Duodenum
    g. Duodenum

21. a. Maltase acts on maltose, a disaccharide.
    b. Peptidase acts on protein segments called peptides.
    c. Secretin stimulates the pancreas to secrete bicarbonate ions.
    d. Enterokinase activates trypsinogen.
    e. Cholecystokinin stimulates contraction of the gallbladder and stimulates the pancreas to secrete digestive enzymes.

22. The presence of chyme is the most important factor that regulates secretions in the small intestine.

23. Teniae coli are three bands of longitudinal muscle in the wall of the large intestine. These three bands comprise the longitudinal muscle layer.

24. The hepatic flexure is the right colonic flexure and is between the ascending colon and transverse colon. The splenic flexure is the left colonic flexure and is between the transverse colon and descending colon.

25. Functions of the large intestine include the absorption of water and electrolytes and the elimination of feces.

26. Falciform ligament.

27. The components of a portal triad are a branch of the hepatic artery, a branch of the hepatic portal vein, and a branch of the hepatic duct. Portal triads are located around the periphery of liver lobules.

28. Hepatic artery and hepatic portal vein.

29. The production of bile.

30. Bilirubin (bile pigment) and cholesterol.

31. Bilirubin is a bile pigment that forms from the decomposition of heme portion of the hemoglobin molecule.

32. The function of the gallbladder is to concentrate and store bile.

33. The pancreas is located along the posterior abdominal wall behind the parietal peritoneum. It extends from the duodenum on the right to the spleen on the left.

34. Pancreatic acinar cells secrete digestive enzymes.

35. a. Amylase acts on polysaccharides to split them into disaccharides.
    b. Trypsin acts on proteins to split them into peptides.
    c. Peptidase enzymes act on peptides to split them into amino acids.
    d. Lipase enzymes split fats into fatty acids and monoglycerides.

36. The presence of acid chyme in the duodenum stimulates the production of the hormone secretin, which stimulates the pancreas to secrete a bicarbonate-rich fluid to neutralize the acid. Proteins and fats in the chyme stimulate the production of the hormone cholecystokinin, which stimulates the pancreas to secrete digestive enzymes.

37. The end products of carbohydrate digestion are the monosaccharides glucose, fructose, and galactose.

38. Pepsin and trypsin act on proteins to split them into peptides. The end products of protein digestion are amino acids.

39. The end products of lipid digestion are free fatty acids and monoglycerides.

40. The blood capillaries in the villi of the small intestine absorb simple sugars and amino acids. The lymph capillaries, called lacteals, also in the villi absorb the fatty acids and monoglycerides.

☞ **Answers to Learning Exercises**

Introduction (Objective 1)

1. G (esophagus)
   H (gallbladder)
   J (large intestine)
   I (liver)
   F (mouth)
   D (pancreas)
   B (pharynx)
   A (salivary gland)
   E (small intestine)
   C (stomach)
   There should be an asterisk by each accessory organ, H, I, D, and A.

Functions of the Digestive System (Objective 2)

1. Mastication
   Hydrolysis
   Chemical digestion
   Deglutition
   Peristalsis
   Absorption
   Defecation

General Structure of the Digestive Tract (Objective 3)

1. A (Innermost layer...)
   B (Contains blood...)
   C (Responsible for most...)
   A (Consists of simple...)
   C (Contains inner ...)
   B (Contains Meissner's...)
   C (Contains the myenteric...)

Regions of the Digestive Tract (Objectives 4-26)

1. Buccinator
   Palate
   Uvula
   Lingual tonsils
   Papillae
   Incisors
   Canines
   Parotid
   Submandibular
   Salivary amylase
   Cleansing action on teeth
   Moistens and lubricates food
   Dissolves molecules for taste
   Begins chemical digestion of carbohydrates

2. G (Alveolar process)
   D (Apical foramen)
   C (Cementum)
   E (Dentin)
   A (Enamel)
   B (Gingiva)
   F (Pulp cavity)
   H (Root canal)

3. Fauces
   Nasopharynx
   Nasopharynx
   Oropharynx
   Uvula
   Epiglottis
   Esophagus
   Esophageal hiatus
   Lower esophageal or cardiac

4. C (Body)
   B (Cardiac region)
   H (Duodenum)
   A (Fundus)
   F (Lower esophageal sphincter)
   G (Pyloric sphincter)
   E (Pylorus)
   D (Rugae)

5. Hydrochloric acid
   Gastrin
   Pepsin
   Chyme
   Absorption of vitamin B_{12}
   Receptivity of duodenum
   Nature of contents (fluidity of chyme)

6. C (Triggered by...)
   A (Begins with thoughts...)
   B (Begins when food...)
   C (Inhibits gastric secretions)
   B (Involves distention...)

7. Duodenum
   Plicae circulares
   Villi
   Microvilli
   Lacteal
   Duodenum
   Ileum
   Enterokinase
   Cholecystokinin
   Secretin
   Lipase
   Cholecystokinin (pancreozymin)
   Chyme in small intestine

8. Absorption of water and electrolytes
   Removal of waste materials

9. H (Where small...)
   L (Three bands...)
   E (Pieces of fat...)
   C (Blind pouch that...)

B (Portion of large...)
G (Right colonic flexure)
K (Left colonic flexure)
M (Portion of large...)
D (Portion of large...)
J (S-shaped curve...)
I (Portion that follows...)
A (Terminal opening...)
F (Series of pouches...)

Accessory Organs of Digestive (Objectives 27-37)

1. Falciform ligament
   Caudate lobe
   Quadrate lobe
   Liver lobule
   Sinusoids
   Hepatic artery
   Hepatic portal vein
   Hepatic vein
   Bile
   Bilirubin
   Emulsify fats
   Gallbladder
   Cystic duct
   Cholecystokinin
   Common bile duct
   Hepatic artery
   Hepatic portal vein
   Hepatic duct
2. Acinar cells
   Pancreatic islets (of Langerhans)
   Pancreatic duct (of Wirsung)
   Pancreatic amylase
   Trypsin
   Lipase
   Secretin
   Cholecystokinin (pancreozymin)

Chemical Digestion (Objectives 38-41)

1. Row 1: Amylase
   Row 2: Maltase, small intestine
   Row 3: Small intestine, sucrose to fructose and glucose
   Row 4: Lactase, small intestine
   Row 5: Pepsin
   Row 6: Pancreas
   Row 7: Peptidase, pancreas
   Row 8: Lipase, pancreas
2. Glucose, fructose, galactose
3. Amino acids
4. Monoglycerides and fatty acids

Absorption (Objective 42)

1. Blood capillary
   Lacteal
   Micelles
   Chylomicron
   Chyle

2. Active transport
   Facilitated diffusion
   Osmosis
   Simple diffusion

☞ Answers to Chapter Self-Quiz

1. b
2. d
3. 32 (permanent) - 20 (deciduous) = 12
4. False
5. B (Opening from...)
   C (Portion of stomach...)
   K (Circular band...)
   J (Circular folds...)
   I (Circular band...)
   E (Curve between...)
   M (Longitudinal folds...)
   O (Longitudinal muscle...)
   A (Shortest part...)
   F (Circular band...)
6. A (Breaks carbohydrates...)
   B (Stimulates secretion...)
   C (Acts on a disaccharide)
   D (Stimulates secretion...)
   E (Breaks proteins...)
   B (Stimulates secretion...)
7. b
8. Production of bile
9. Common bile duct and pancreatic duct
10. C B (glucose)
    P B (amino acids)
    F L (monoglycerides)
    C B (fructose)
    F L (fatty acids)
    C B (galactose)

☞ Answers to Terminology Exercises

Amylase
Anorexia
Enteritis
Gastralgia
Sublingual

Pertaining to bile
Inflammation of the gallbladder
Toothache
Inflammation of the liver
Inflammation of salivary glands

D (after a meal)
A (surgical excision of the gallbladder)
C (vomiting blood)
E (pain in the rectum and anus)
B (inflammation of the gums)

☞ Answers to Fun and Games

1. Gingiva
2. Volvulus
3. Frenulum
4. Duodenum
5. Dysphagia
6. Pepsinogen
7. Deglutition
8. Peristalsis
9. Mastication
10. Borborygmus

☞ Quiz/Test Questions

Note: There are fifty multiple-choice questions for this chapter in the computerized test bank.

Name the following:

1. Nerve plexus in the muscle layer of the gastrointestinal wall.
   Answer: myenteric (Auerbach's) plexus.

2. Type of muscle tissue in the tongue.
   Answer: skeletal muscle.

3. Structure that separates oral cavity from nasal cavity.
   Answer: palate.

4. Largest of the salivary glands.
   Answer: parotid gland.

5. Enzyme in saliva.
   Answer: amylase.

6. Structure that controls passage of food from esophagus into stomach.
   Answer: lower esophageal sphincter (cardiac sphincter).

7. Longitudinal folds in mucosa of stomach.
   Answer: rugae.

8. Enzyme secreted by stomach cells.
   Answer: pepsinogen (inactive form) or pepsin (active form).

9. First, or proximal, part of the small intestine.
   Answer: duodenum.

10. Lymph capillaries that absorb monoglycerides and fat soluble vitamins.
    Answer: lacteals.

11. Hormone that stimulates release of bicarbonate-rich fluid from the pancreas.
    Answer: secretin.

12. Components of a portal triad in the liver.
    Answer: branch of portal vein, branch of hepatic artery, branch of hepatic duct.

13. Digestive function of the liver.
    Answer: production of bile.

14. Where the pancreatic duct empties pancreatic exocrine secretions.
    Answer: duodenum.

15. Three monosaccharides that are the end products of carbohydrate digestion.
    Answer: glucose, fructose, galactose.

True/False Questions:

1. Amino acids are converted back to proteins in the cells of the microvilli before they are absorbed into the blood.
   Answer: False; triglycerides are reformed in the microvilli cells before they are absorbed into the lacteals.

2. The complete deciduous dentition has 20 teeth and the permanent dentition has 32 teeth.
   Answer: True.

3. Impulses along the vagus nerve stimulate gastric secretions in the cephalic phase of the regulation of gastric secretions.
   Answer: True.

4. The passage of chyme through the cardiac sphincter initiates the intestinal phase of gastric secretions.
   Answer: False; it is the pyloric sphincter.

5. The splenic flexure is between the ascending colon and transverse colon.
   Answer: False; it is the hepatic flexure.

6. The falciform ligament separates the right and left lobes of the liver on the anterior surface and attaches the liver to the anterior abdominal wall.
   Answer: True.

7. The gallbladder is attached to the visceral surface of the liver by the common bile duct.
   Answer: False; it is the cystic duct.

8. Secretin stimulates the pancreas to secrete digestive enzymes.
   Answer: False; cholecystokinin stimulates the digestive enzymes.

9. Pepsin and trypsin both function in the digestion of proteins.
   Answer: True.

10. Bile salts emulsify fats prior to digestion and then aid in the absorption of the end products of fat digestion.
    Answer: True.

17 Metabolism and Nutrition

Key Terms/Concepts

Acetyl CoA A molecule formed from pyruvic acid in the mitochondria when oxygen is present; a key molecule in the metabolism of carbohydrates, proteins, and lipids.

Anabolism Building up, or synthesis, reactions that require energy and make complex molecules out of two or more smaller ones; opposite of catabolism.

Basal metabolic rate Amount of energy that is necessary to maintain life and keep the body functioning at a minimal level.

Beta oxidation Catabolic reaction in which a 2-carbon segment is removed from a fatty acid and is converted to acetyl CoA.

Catabolism Reactions that break down large molecules into two or more smaller ones with the release of energy; opposite of anabolism.

Citric acid cycle Aerobic series of reactions that follows glycolysis in glucose metabolism to release energy and carbon dioxide; also called Kreb's cycle .

Complete protein A protein that contains all of the essential amino acids.

Core temperature The temperature deep in the body; temperature of the internal organs.

Deamination Catabolic reaction in which an amino group is removed from an amino acid to form ammonia and a keto acid; occurs in the liver as part of protein catabolism.

Gluconeogenesis Process of forming glucose from noncarbohydrate nutrient sources such as proteins and lipids.

Glycogenesis Series of reactions that converts glucose or other monosaccharides into glycogen for storage.

Glycogenolysis Series of reactions that converts glycogen into glucose.

Glycolysis Anaerobic series of reactions that produces 2 molecules of pyruvic acid from one molecule of glucose; first series of reactions in the catabolism of glucose.

Lipogenesis Series of reactions in which lipids are formed from other nutrients.

Minerals Inorganic substances that are needed in minute amounts in the diet to maintain growth and good health, but do not supply energy.

Nutrition Science that studies the relationship of food to the functioning of the living organism; acquisition, assimilation, and utilization of nutrients contained in food.

Thermogenesis Production of heat in response to food intake.

Vitamins Organic compounds that are needed in minute amounts to maintain growth and good health; do not supply energy but are necessary to release energy from carbohydrates, proteins, and lipids.

Chapter Objectives

Upon completion of this chapter the student should be able to:
1. Define the terms metabolism and nutrition.
2. Distinguish between anabolism and catabolism.
3. Name the molecule that represents stored chemical energy in the cell.
4. Use the terms glucose, anaerobic, cytoplasm, pyruvic acid, and ATP to describe glycolysis.
5. Use the terms pyruvic acid, acetyl CoA, citric acid cycle, mitochondria, and ATP to describe the aerobic phase of cellular respiration.
6. Define glycogen, glycogenesis, glycogenolysis, and gluconeogenesis.
7. List six uses of proteins in the body.
8. Explain how amino acids can be used for energy if there is insufficient carbohydrate or fat.
9. Describe the pathway by which fatty acids are broken down to produce ATP.
10. Name two key molecules in the metabolism and interconversion of carbohydrates, proteins, and fats.
11. Define the term Calorie.
12. List three uses of energy in the body.
13. Explain what is meant by basal metabolism and state four factors that influence it.
14. State the avenue of energy use that can be controlled voluntarily.
15. Define thermogenesis.

16. List three functions of carbohydrates in the body.
17. State the number of Calories in one gram of pure carbohydrate.
18. Distinguish between simple sugars and complex carbohydrates.
19. Name three types of complex carbohydrates that are important in nutrition and give a source for each one.
20. Explain the importance of fiber in the diet.
21. State three categories of protein function in the body.
22. State the number of Calories in one gram of protein.
23. Distinguish between essential and nonessential amino acids and between complete and incomplete proteins.
24. Explain why two or more incomplete proteins should be eaten in the same meal.
25. List six functions of fats in the body.
26. State the number of Calories in one gram of fat.
27. Discuss the American Heart Association's recommendations for dietary intake of fats.
28. Identify vitamins as either water soluble or fat soluble.
29. Explain why vitamins are important in the diet.
30. List four uses of minerals in the body.
31. State five reasons water is necessary in the body.
32. Distinguish between core temperature and shell temperature.
33. Explain three ways in which core temperature is maintained when the environmental temperature is cold.
34. State four ways in which heat is lost from the body.
35. Describe the general mechanism by which the body maintains core temperature and identify the region of the brain that integrates this mechanism.

☞ Chapter Outline/Summary

Introduction (Objective 1)
- Metabolism is the sum of all the chemical reactions in the body. Within a cell, the reactions are called cellular metabolism. Enzymes speed up the reactions.
- Nutrition is the acquisition, assimilation, and utilization of the nutrients that are contained in food. The utilization of the nutrients is a part of metabolism.

Metabolism of Absorbed Nutrients (Objectives 2-15)

Anabolism (Objective 2)
- Anabolism uses energy to build large molecules from smaller ones.
- Dehydration synthesis is a type of anabolic reaction in which a water molecule is removed when a larger molecule is synthesized from two smaller ones.

Catabolism (Objective 2)
- Catabolism releases energy when large molecules break down into smaller ones.
- Catabolic reactions within the cell that release energy for use by the cell are termed cellular respiration.

Energy from foods (Objectives 3-10)
- Cellular respiration utilizes the absorbed end products of digestion and stores the energy in the high energy bonds of adenosine triphosphate (ATP).
- Carbohydrates (Objectives 4-6)
  - The first step in the catabolism of glucose is glycolysis, which takes place in the cytoplasm and is anaerobic. Two pyruvic acid molecules are produced from one glucose and there is a net gain of 2 ATP.
  - If oxygen is present, pyruvic acid enters the mitochondria for the aerobic phase of cellular respiration. The pyruvic acid is incorporated into acetyl CoA, which enters the citric acid cycle. The complete breakdown of glucose produces 36-38 ATP.
  - Glycogen is the storage form of glucose. Glycogen is synthesized from glucose by glycogenesis. Glycogenolysis is the breakdown of glycogen into glucose and gluconeogenesis is the production of glucose from noncarbohydrate sources.
- Proteins (Objectives 7 and 8)
  - The end products of protein digestion are amino acids.
  - Amino acids are used to synthesize proteins to build new tissues and replace damaged tissues. They are also used to synthesize hemoglobin, hormones, enzymes, and plasma proteins.
  - Amino acids may be used as an energy source by removing the amino group (deamination). The resulting keto acid enters the citric acid cycle to produce energy.
- Lipids (Objective 9)
  - The end products of lipid digestion are monoglycerides and fatty acids, which are an important source of energy.

○ Fatty acids are catabolized by beta oxidation, which removes 2-carbon segments from fatty acid chains and converts the segments to acetyl CoA. The acetyl CoA enters the citric acid cycle to produce ATP.

- Interconversion of Carbohydrates, Proteins, and Lipids (Objective 10)
  ○ The end products of carbohydrate, protein, and lipid metabolism can be interconverted when necessary.
  ○ Pyruvic acid and acetyl CoA are key molecules in the metabolism and interconversion of carbohydrates, proteins, and fats.

Uses for Energy (Objectives 11-15)
- Energy from food is measured in Calories. One Calorie is the amount of energy required to raise the temperature of one kilogram of water from 14° Celsius to 15° Celsius.
- Basal metabolism (Objectives 12 and 13)
  ○ Basal metabolism is the energy that is necessary to keep the body functioning at a minimal level.
  ○ It is influenced by sex, muscle mass, age, hormones, and body temperature.
- Physical activity (Objectives 12 and 14)
  ○ Increasing physical activity increases energy expenditure in the body.
  ○ Physical activity is the only means of voluntarily controlling energy expenditure.
- Thermogenesis (Objectives 12 and 15)
  ○ Thermogenesis is the production of heat through the assimilation of food.
  ○ Digestion, absorption, and metabolism use chemical energy and heat energy is produced.

Basic Elements of Nutrition (Objectives 16-31)
Carbohydrates (Objectives 16-20)
- Carbohydrates provide energy, add bulk to the diet, and are used to synthesize other compounds.
- One gram of pure carbohydrate yields 4 Calories of energy.
- Sugars include the monosaccharides and disaccharides. Complex polysaccharides are formed from long chains of glucose and include starch, glycogen, and fiber.
- Starch is the storage form of glucose in plants, glycogen is the storage form in animals. Fiber comes from plant sources and is not digestible.
- Fiber adds bulk to the diet, enhances absorption of nutrients, and helps move the feces in

the large intestine. Many diseases are related to decreased dietary fiber.

Proteins (Objectives 21-24)
- Proteins provide structure, regulate body processes, and provide energy.
- One gram of protein yields 4 Calories of energy.
- Essential amino acids must be supplied in the diet; they cannot be synthesized in the body. Nonessential amino acids can be synthesized from other amino acids that are available.
- Complete proteins contain all the essential amino acids. Incomplete proteins lack one or more of the essential amino acids. Incomplete proteins should be eaten in combinations that provide all the essential amino acids.

Lipids (Objectives 25-27)
- Lipids are an important source of energy, provide essential fatty acids, transport vitamins, are components of certain structural elements, provide heat insulation, and form protective cushions.
- One gram of fat yields 9 Calories of energy.
- Although fats are essential in the diet, most Americans eat more than necessary and they eat too many saturated fats. The American Heart Association recommends that less than 75 grams of fat be eaten every day and that less than 25 grams of this be in the form of saturated fats.

Vitamins (Objectives 28 and 29)
- Vitamins are organic molecules that are necessary for good health.
- Vitamins A, D, E, and K are fat soluble vitamins. The other vitamins are water soluble.
- Many vitamins are part of enzyme molecules, which are incomplete without the vitamin portion. These enzymes are necessary for the chemical reactions throughout the body.

Minerals (Objective 30)
- Minerals are inorganic substances that are necessary in small amounts to maintain good health.
- Minerals are obtained from plants since the plants absorb them from the ground.
- Some minerals are components of body structures; some are parts of enzyme molecules; others help control fluid levels; and others become part of larger organic molecules.

Water (Objective 31)
- Water is an essential component of the diet. The average adult requires about 2.5 liters of water every day.

- Water is an integral part of body cells, provides a medium for chemical reactions, is a transport medium, is a lubricant, and helps maintain body temperature.

Body Temperature (Objectives 32-35)

- Core temperature is the temperature of the internal organs.
- Shell temperature is the temperature at the body surface.
- Heat is produced by the catabolism of nutrients. The body produces additional heat by muscular contraction (shivering) and increasing metabolic rate (hormones). Heat is conserved by constriction of cutaneous blood vessels, which keeps the blood warmer by keeping it away from the cold surface of the body.
- Heat is lost from the body through radiation, conduction, convection, and evaporation.
- Body temperature is regulated by a homeostatic negative feedback mechanism that is integrated by the hypothalamus of the brain.

☞ **Answers to Review Questions**

1. Metabolism includes all the chemical reactions in the body. Nutrition is the acquisition, assimilation, and utilization of nutrients. One aspect of metabolism is the utilization of nutrients.
2. Catabolism releases chemical energy and this is usually in the form of adenosine triphosphate (ATP).
3. Glycolysis takes place in the cytoplasm and the end products are pyruvic acid and a net gain of 2 ATP.
4. Pyruvic acid is first converted to acetyl-CoA, which enters the citric acid cycle to be catabolized into carbon dioxide and energy. The aerobic processes take place in the mitochondria.
5. When blood glucose levels are high, the excess glucose is converted to glycogen by the process of glycogenesis. When blood glucose levels begin to decline the stored glycogen is broken down to glucose by the process of glycogenolysis.
6. Amino acids are used to synthesize the wide assortment of proteins needed by the body.
7. Amino acids are prepared for used as an energy source by the process of deamination. In this process, the amino group is removed, which leaves a keto acid that can be used to provide energy.
8. Beta oxidation is the chemical reaction involved in the catabolism of fatty acids. This process splits the fatty acids into 2-carbon segments, which become acetyl CoA.
9. The two key molecules in gluconeogenesis are acetyl CoA and pyruvic acid.
10. The total metabolic rate includes the energy expended to maintain life at a minimal level (basal metabolism) plus the energy for physical activity plus the energy required for thermogenesis.
11. Physical activity.
12. Glucose.
13. Carbohydrate = 4C/gram
    Protein = 4C/gram
    Fat = 9C/gram
14. Fructose: honey and many fruits
    Sucrose: table sugar
    Lactose: milk
    Maltose: sprouting grains, corn syrup
15. Starch is the storage form of energy in plants and can be digested by humans to provide energy. Fiber is a complex polysaccharide that cannot be digested by the enzymes in the human digestive system.
16. Hormones and enzymes are proteins that regulate body functions and control chemical reactions.
17. Certain amino acids are termed "essential" because they cannot be synthesized in the body and must be supplied in the diet.
18. One or more essential amino acids are lacking in an incomplete protein.
19. Fats.
20. The American Heart Association recommends that less than 10% of the daily caloric intake, or about 25 grams, be in the form of saturated fats. Animal fats are high in saturated fats. This includes most meats, eggs, butter, and whole-milk products.
21. Vitamins A, D, E, and K, which are the fat soluble vitamins, require fats for absorption and utilization.
22. Vitamins and minerals do not supply energy, but they are necessary in the diet because they are part of the enzymes that work to release the energy from the carbohydrates, proteins, and fats. They are also necessary in many other aspects of the body such as DNA synthesis, blood clotting, muscle contraction, hemoglobin synthesis, and bone structure.
23. Fruits and vegetables are an important source of minerals in the diet.
24. The body is about 60% water by weight and it is essential that water balance be maintained. Water is an integral part of cells, is a lubricant, helps maintain body temperature,

is a transport medium, and provides an appropriate medium for chemical reactions.

25. Homeotherms are warm-blooded animals. It means they have the ability to maintain a constant internal body temperature even when there are variations in external temperature.

26. Core temperature is about 1°C greater than shell temperature.

27. The main source of heat to maintain core temperature is the catabolism of nutrients. About 60% of the energy in nutrients is released as heat.

28. Most of the heat loss from the body is by radiation. This accounts for about 60% of the heat loss.

29. The body's "thermostat" is located in the hypothalamus.

☞ Answers to Learning Exercises

Introduction (Objective 1)
1. Metabolism
   Cellular metabolism
   Enzymes
   Nutrition

Metabolism of Absorbed Nutrients (Objectives 2-15)
1. H (Most important simple sugar...)
   C (Term that means...)
   L (Reactions in which glucose is...)
   M (Fate of pyruvic acid...)
   B (Term that means...)
   A (Molecule that enters...)
   I (Storage form of glucose)
   E (Location of glycolysis reactions)
   O (Location of citric...)
   J (Conversion of glucose to glycogen)
   N (Conversion of glucose to fat)
   K (Conversion of glycogen into glucose)
   G (Conversion of non-carbohydrate...)
   F (Principal reaction...)
   D (Reactions that convert...)

2. Build new tissues
   Replace damaged tissues
   Synthesis of hemoglobin
   Synthesis of enzymes
   Synthesis of hormones
   Synthesis of plasma proteins

3. Acetyl CoA
   Pyruvic acid

4. A Calorie is the amount of heat required to raise the temperature of one kilogram of water one degree Celsius, from 14° to 15°.

5. Basal metabolism
   Physical activity
   Thermogenesis

Basal metabolism accounts for most of the energy used and physical activity is the one that can be controlled voluntarily.

Basic Elements of Nutrition (Objectives 16-31)
1. C (Primary energy source)
   P (Regulate body processes)
   L (Major component of cell membranes)
   L (Transport vitamins A, D, E, K)
   C (Add bulk to the diet)
   L (Fatty acids)
   L (Provide insulation and protection)
   P (Provide structure)
   C (Glucose)
   L (Concentrated energy)
   C (Fiber)
   P (Hormones, enzymes)
   L (Steroids)
   C (Glycogen)
   L (Triglycerides)
   P (Amino acids)

2. 4 (1 gram pure carbohydrate)
   4 (1 gram pure protein)
   9 (1 gram pure fat)
   20 (5 grams carbohydrate)
   16 (4 grams protein)
   27 (3 grams fat)

3. Amino acids that cannot be synthesized in the body and must be supplied in the diet are called **essential** amino acids. A protein that contains all of these amino acids is called a **complete** protein. Other proteins, called incomplete proteins should be eaten in combinations to provide all of the **essential** amino acids.

4. The American Heart Association recommends that no more than **30%** of the daily calorie intake should be in the form of fats. Further, they recommend a reduction in the amount of saturated fats in the diet. In general, foods that are high in saturated fats are also high in **cholesterol**. Cholesterol intake should be limited to less than **250 mg** per day.

5. W (Thiamine)
   W (Niacin)
   F (Vitamin A)
   W (Vitamin C)
   F (Vitamin D)
   W (Riboflavin)

6. W (Medium for chemical reactions)
   V (Release energy from nutrients)
   V (Nucleic acid synthesis)
   M (Incorporated into bones and teeth)
   W (Maintenance of body temperature)
   W (Regulation of body fluid levels)

Body Temperature (Objectives 32-35)

1. E (Byproduct …)
   F (Location…)
   C (Temperature…)
   D (Heat…)
   A (Distributes…)
   J (Two methods…)
   B
   H (Two methods…)
   G
   I (Temperature…)

☞ **Answers to Chapter Self-Quiz**

1. C (releases energy)
   C (cellular respiration)
   A (dehydration synthesis)
   C (breaks large molecules into smaller ones)
2. b
3. c
4. Glycogen
5. D (a reaction in protein catabolism)
   B (breaks off…)
   D (produces ammonia)
   D (a reaction in…)
   B (produces acetyl CoA)
   D (a step in the…)
   B (a reaction in…)
6. Energy used for basal metabolism
7. 8 gm x 4C/gm = 32 Calories from CHO
   32C/9C/gm = 3.56 grams of fat
8. B (a monosaccharide)
   A (nondigestible complex…)
   C (formed from…)
   D (glucose storage…)
   E (table sugar)
9. C (collagen synthesis)
   A (formation of pigments in retina)
   K (synthesis of clotting factors)
   B_{12} (formation of erythrocytes)
   Folic acid (synthesis of DNA)
10. d

☞ **Answers to Terminology Exercises**

Pyrogenesis
Glycogenolysis
Nutrition
Gluconeogenesis
Hyperthermia

To take apart with water
Low or deficient temperature
To form lipids
Inhibits fever production
Producing heat

E (a substance required for life)
D (poor nourishment)
A (throwing down or taking apart)
C (taking apart fats)
B (process of removing an amino group

☞ **Answers to Fun and Games**

1. Catabolism (14 points)
   Anabolism (12 points)
2. Glycolysis (19 points)
   Gluconeogenesis (21 points)
3. Glucose (11 points)
   Fructose (11 points)
   Galactose (13 points)
4. Sucrose (8 points)
   Maltose (9 points)
   Lactose (9 points)
5. Glycogen (16 points)
   Starch (9 points)
   Fiber (8 points)
6. Triglyceride (19 points)
   Steroid (8 points)
   Cholesterol (16 points)
7. Radiation (10 points)
   Conduction (13 points)
   Convection (15 points)
   Evaporation (16 points)
8. Calcium (11 points)
   Chloride (13 points)
   Phosphorus (18 points)
   Sodium (8 points)
   Potassium (12 points)
   Magnesium (13 points)
   Iron (4 points)
   Iodine (7 points)
   Zinc (9 points)
   Fluoride (12 points)
9. Thiamine (11 points)
   Riboflavin (17 points)
   Niacin (7 points)
   Pyridoxine (19 points)
   Cyanocobalamin (21 points)
10. Deamination (13 points)

☞ **Quiz/Test Questions**

Note: There are fifty multiple-choice questions for this chapter in the computerized test bank.

Name the following:

1. Reactions that break large molecules into smaller ones and release energy.
   Answer: catabolism.

2. Molecule that stores energy in a form that is usable by the cell.
   Answer: adenosine triphosphate (ATP).

3. Anaerobic catabolism of glucose.
   Answer: glycolysis.

4. End product of glycolysis if oxygen is present.
   Answer: pyruvic acid.

5. Where in the cell the reactions of the citric acid cycle occur.
   Answer: in the mitochondria.

6. Storage form of glucose in the human body.
   Answer: glycogen.

7. Process by which noncarbohydrate nutrient sources are converted to glucose.
   Answer: gluconeogenesis.

8. Principal reaction by which proteins are prepared for use as an energy source.
   Answer: deamination.

9. Process by which segments of fatty acids are converted to acetyl Co-A.
   Answer: beta oxidation

10. Energy required to keep body functioning at a minimal level.
    Answer: basal metabolic rate.

11. Complex polysaccharides that cannot be digested in the human body.
    Answer: fiber.

12. Amino acids that are needed by the body but cannot be synthesized in the human body.
    Answer: essential amino acids.

13. Lipid that generally is in the same food source as saturated fats.
    Answer: cholesterol.

14. Inorganic substances that are needed by the body in small amounts.
    Answer: minerals.

15. Mechanism that accounts for about 60 percent of the body's heat loss.
    Answer: radiation.

True/False Questions:

1. Proteins provide more energy per gram that do carbohydrates.
   Answer: False; they both provide 4C/g.

2. Vitamins C and K are fat soluble vitamins.
   Answer: False; vitamin C is water soluble.

3. More energy is produced by aerobic catabolism of glucose than by anaerobic catabolism of glucose.
   Answer: True.

4. If oxygen is present, pyruvic acid is converted to lactic acid, which enters the citric acid cycle.
   Answer: False; pyruvic acid is converted to acetyl CoA.

5. Glucose may be converted to fat by the process of lipogenesis.
   Answer: True.

6. Deamination results in the formation of ammonia, which is then converted to urea.
   Answer: True.

7. The conversion of proteins to fats is called gluconeogenesis.
   Answer: False; it is called lipogenesis, proteins to carbohydrates is gluconeogenesis.

8. Fiber and starch, both complex polysaccharides, provide the same number of calories per gram.
   Answer: False; fiber provides no calories because it is not digestible.

9. Four grams of fat yields the same number of calories as 9 grams of carbohydrate.
   Answer: True.

10. Fats should be avoided in the diet.
    Answer: False; fats are essential in the diet, however, intake should be limited to meet healthy standards.

18 Urinary System and Body Fluids

☞ Key Terms/Concepts

Acidosis Condition in which the blood has a lower pH than normal.

Alkalosis Condition in which the blood has a higher pH than normal.

Detrusor muscle The smooth muscle in the wall of the urinary bladder.

Glomerular capsule Double layered epithelial cup that surrounds the glomerulus in a nephron; also called Bowman's capsule.

Glomerulus Cluster of capillaries in the nephron through which blood is filtered.

Interstitial fluid Portion of the extracellular fluid that is found in the tissue spaces.

Intracellular fluid (ICF) The fluid inside body cells.

Intravascular fluid Portion of extracellular fluid that is in the blood; plasma.

Juxtaglomerular apparatus Complex of modified cells in the afferent arteriole and the ascending limb/distal tubule in the kidney, which helps regulate blood pressure by secreting renin; consists of the macula densa and juxtaglomerular cells.

Micturition Act of expelling urine from the bladder; also called urination or voiding.

Nephron Functional unit of the kidney consisting of a renal corpuscle and a renal tubule.

Renal cortex Outer portion of the kidney that appears granular.

Renal pelvis Large cavity in the central region of a kidney that collects the urine as it is produced.

Renal pyramids Triangular shaped regions in the kidney that appear striated.

Renal medulla Inner portion of the kidney consisting of renal pyramids.

Renal tubule Tubular portion of the nephron that carries the filtrate away from the glomerular capsule and where tubular reabsorption and secretion occurs.

☞ Chapter Objectives

Upon completion of this chapter the student should be able to:

1. State six functions of the urinary system.
2. Describe the location of the kidneys.
3. Label the capsule, cortex, renal pyramids, renal columns, papillae, major and minor calyces, renal pelvis, and ureter on an illustration of a frontal section through the kidney.
4. Distinguish between the renal corpuscle and the renal tubule.
5. Draw a diagrammatic representation of a nephron and label the glomerulus, glomerular capsule, proximal convoluted tubule, loop of the nephron with descending and ascending limbs, and the distal convoluted tubule.
6. Identify the parts of a nephron that are in the cortex and those that are in the medulla.
7. Name the structures that collect filtrate from the nephrons and takes it to a minor calyx. Indicate where these structures are located in the kidney.
8. Name the two parts of the juxtaglomerular apparatus and state where they are located.
9. Trace the pathway of blood flow through the kidney from the renal artery to the renal vein.
10. Name the structure that transports urine away from the kidney.
11. Describe the location of the urinary bladder.
12. Give the name of the folds of the mucosa and the name of the smooth muscle in the wall of the urinary bladder.
13. Name the openings that form the three points of the trigone.
14. Define the term micturition.
15. Distinguish between the internal and external urethral sphincters with respect to location and type of muscle.
16. Compare the length of the female urethra with the length of the male urethra.
17. List the three basic steps in urine formation.
18. State the direction of fluid flow in glomerular filtration.
19. Identify three different types of pressure that affect the rate of glomerular filtration and describe how these interact.
20. State the direction substances move in tubular reabsorption.

21. Distinguish between the mechanism for solute reabsorption and the mechanism for water reabsorption.
22. Explain why some substances, such as glucose, have limited reabsorption and what happens when concentration exceeds this limit.
23. Identify the portion of the nephron tubule that is not permeable to water.
24. Define tubular secretion and name four substances that are added to the urine during this process.
25. Explain how urine production has a role in maintaining blood concentration and volume.
26. Name two hormones that affect kidney function and explain the effect of each one.
27. Name the enzyme that stimulates the production of angiotensin II and is produced by the kidneys.
28. Describe two mechanisms by which angiotensin II increases blood pressure.
29. State the color, specific gravity, and pH of urine.
30. Describe the chemical composition of urine.
31. Name at least four abnormal constituents of urine that may indicate pathologic conditions.
32. Compare the volume of intracellular fluid with the volume of extracellular fluid and the volume of interstitial fluid with the volume of intravascular fluid.
33. List three sources of fluid intake.
34. List four avenues of fluid loss from the body.
35. Identify three major ions in the extracellular fluid and two major ions in the intracellular fluid.
36. Identify the primary hormone that regulates electrolyte concentration.
37. State the normal pH range of the blood and the terms used to indicate deviations from the normal.
38. Describe three mechanisms for maintaining normal blood pH.

☞ **Chapter Outline/Summary**

Functions of the Urinary System (Objective 1)
- The urinary system rids the body of waste materials, regulates fluid volume, maintains electrolyte concentrations in body fluids, controls blood pH, secretes erythropoietin, and secretes renin.

Components of the Urinary System (Objectives 2-16)
- The components of the urinary system are the kidneys, ureters, urinary bladder, and urethra.

Kidneys (Objectives 2-9)
- Location (Objective 2)
  - The kidneys are retroperitoneal against the posterior abdominal wall.
  - They are between the levels of the twelfth thoracic and third lumbar vertebrae.
- Macroscopic structure (Objective 3)
  - An indentation called the hilus leads to the renal sinus. The renal artery, renal vein, and ureter penetrate the kidney at the hilus.
  - The kidney is enclosed by a capsule and surrounded by perirenal fat.
  - Internally, the renal cortex, renal medulla, pyramids, renal columns, papillae, calyces, and pelvis are visible.
- Nephrons (Objectives 4-6)
  - The nephron is the functional unit of the kidney.
  - A nephron consists of a renal corpuscle and a renal tubule.
  - The renal corpuscle consists of a glomerulus and a glomerular capsule.
  - The renal tubule consists of a proximal convoluted tubule, nephron loop with descending and ascending limbs, and distal convoluted tubule.
  - The nephron loop is the only part of a nephron located in the pyramids. The other portions are in the cortex.
- Collecting ducts (Objective 7)
  - Urine passes from the nephrons into collecting ducts, then into the minor calyces.
  - Collecting ducts are located in the pyramids.
- Juxtaglomerular apparatus (Objective 8)
  - The juxtaglomerular apparatus consists of modified cells where the ascending limb of the nephron loop (macula densa) comes in contact with the afferent arteriole (juxtaglomerular cells).
  - The juxtaglomerular cells secrete renin.
- Blood flow through the kidney (Objective 9)
  - Blood flows through the kidney in the sequence renal artery -> segmental arteries -> interlobar arteries -> arcuate arteries -> interlobular arteries -> afferent arteriole -> glomerulus -> efferent arteriole -> peritubular capillaries.
  - From the capillaries back to the renal vein, the sequence is peritubular capillaries -> interlobular veins -> arcuate veins -> interlobar veins -> segmental veins -> renal vein.

Ureters (Objective 10)

- The ureter transports urine from the kidney to the urinary bladder.
- The ureter is continuous with the renal pelvis.
- As it leaves the kidney, the ureter descends along the posterior abdominal wall and is retroperitoneal.
- The ureters enter the urinary bladder on the posterior and inferior surface.

Urinary bladder (Objectives 11-14)

- The urinary bladder is posterior to the symphysis pubis and below the parietal peritoneum in the pelvic cavity.
- The lining of the urinary bladder is a mucous membrane with folds called rugae. The smooth muscle in the wall is the detrusor muscle.
- The trigone, in the floor of the urinary bladder, is outlined by the two ureters and the internal urethral orifice.
- Micturition is the act of expelling urine from the bladder.

Urethra (Objectives 15 and 16)

- The urethra transports urine from the bladder to the exterior of the body.
- The flow is controlled by two sphincters. The internal sphincter is located where the urethra leaves the bladder and is smooth muscle. The external sphincter is located where the urethra penetrates the pelvic floor and is skeletal muscle.
- In females the urethra is short. In males it is longer and extends the length of the penis. The male urethra is divided into the prostatic urethra, membranous urethra, and the spongy urethra.
- The urethra opens to the exterior through the external urethral orifice.

Urine Formation (Objectives 17-28)

- The work of the kidneys is accomplished through the formation of urine, which involves glomerular filtration, tubular reabsorption, and tubular secretion.

Glomerular filtration (Objectives 17-19)

- In glomerular filtration, plasma components cross the filtration membrane from the glomerulus into the glomerular capsule.
- The rate of glomerular filtration depends on the net filtration pressure. This pressure is the result of the interaction of the blood pressure in the glomerulus, hydrostatic pressure in the capsule, and osmotic pressure in the blood.

Tubular reabsorption (Objectives 20-23)

- Tubular reabsorption moves substances from the filtrate into the blood in the capillaries and reduces the volume of urine.
- Most of the solutes are reabsorbed by active transport mechanisms.
- Reabsorption of some solutes is limited by carrier molecules. When the concentration of these solutes exceeds renal threshold, the excess appears in the urine.
- Water is reabsorbed by osmosis in all parts of the tubule except the ascending limb of the nephron loop, which is impermeable to water.

Tubular secretion (Objective 24)

- Tubular secretion adds substances to the urine.
- Hydrogen ions, potassium ions, creatinine, histamine, and penicillin are examples of substances that are added to the urine by tubular secretion.

Regulation of urine concentration and volume (Objectives 25-28)

- By altering the concentration and volume of urine, the kidneys have a major role in maintaining blood concentration, volume, and pressure.
- Aldosterone increases sodium reabsorption in the kidney tubules. This causes sodium retention, and secondarily, water retention. This will also increase blood volume and blood pressure.
- Antidiuretic hormone increases water reabsorption by the kidney tubules. This decreases urine volume and adds to the fluid in the body. This will also increase blood volume and blood pressure.
- Renin, which promotes the production of angiotensin II, is produced by the juxtaglomerular cells.
- Angiotensin II is a vasoconstrictor, which increases blood pressure. It also stimulates the release of aldosterone, which acts on the kidney tubules to conserve sodium and water. This also increases blood volume and blood pressure.

Characteristics of Urine (Objectives 29-31)

- Freshly voided urine has a clear yellow color, a specific gravity of 1.001 to 1.035, and a pH between 4.6 and 8.0.
- Urine is about 95% water and 5% solutes.
- Abnormal constituents of urine include albumin, glucose, blood cells, ketone bodies, and microbes.

Body Fluids (Objectives 32-38)
Fluid compartments (Objective 32)
- Fluids make up 60% of the adult body weight.
- Intracellular fluid is the fluid inside the body cells and it accounts for 2/3 of the total body fluid.
- Extracellular fluid is the fluid outside the body cells and it accounts for 1/3 of the total body fluid.
- Extracellular fluid is further divided into intravascular fluid (1/5) and interstitial fluid (4/5).

Intake and output of fluid (Objectives 33 and 34)
- Normally fluid intake equals output. This should be about 2500 ml (2.5 L) per day.
- Sources of fluid intake are beverages, food, and metabolic water.
- Avenues of fluid loss are through the kidneys, skin, lungs, and gastrointestinal tract.

Electrolyte balance (Objectives 35 and 36)
- Electrolyte concentrations in the different fluid compartments are different, but they remain relatively constant within each compartment.
- Sodium (Na^+), chloride (Cl^-), and bicarbonate (HCO_3^-) ions are the predominant ions in the extracellular fluid.
- Potassium (K^+) and phosphates ($H_2PO_4^-$ and HPO_4^{-2}) are the predominant ions in the intracellular fluid.
- Aldosterone is the primary regulator of electrolyte concentration through reabsorption of sodium and potassium.

Acid-base balance (Objectives 37 and 38)
- The normal pH of the blood ranges between 7.35 and 7.45.
- Deviations below normal are called acidosis and deviations above normal are called alkalosis.
- Acid-base balance is maintained through the action of buffers, the lungs, and the kidneys.

☞ **Answers to Review Questions**

1. The principal function of the urinary system is to maintain the volume and composition of body fluids within normal limits.
2. Kidneys, ureters, urinary bladder, and urethra.
3. The kidneys are retroperitoneal and located between the twelfth thoracic and third lumbar vertebrae.
4. See Figure 18-2 in the textbook.
5. The renal corpuscle consists of a glomerulus and a glomerular (Bowman's) capsule. The renal tubule consists of a proximal convoluted tubule, nephron (Henle's) loop, and distal convoluted tubule.
6. Glomerular capsule → proximal convoluted tubule → nephron loop → distal convoluted tubule → collecting duct → minor calyx → major calyx → renal pelvis → ureter.
7. Regions of the afferent arteriole and ascending limb of the nephron loop are modified to form the juxtaglomerular apparatus, which secretes renin.
8. Afferent arterioles are branches of the interlobular arteries.
9. Ureters transport urine from the kidneys to the urinary bladder and enter the bladder on the posterior inferior surface.
10. The detrusor muscle is the smooth muscle in the wall of the urinary bladder.
11. The trigone is a triangular area in the floor of the urinary bladder formed by the two openings for the ureters and the single opening for the urethra.
12. The internal urethral sphincter is visceral, or smooth and involuntary, muscle. The external urethral sphincter is skeletal, or striated and voluntary, muscle.
13. Prostatic urethra, membranous urethra, and spongy urethra.
14. Glomerular filtration, tubular reabsorption, and tubular secretion.
15. 180 liters/day.
16. Blood cells and protein molecules.
17. Glomerular osmotic pressure and capsular hydrostatic pressure are inversely proportional to net filtration pressure.
18. Most (65%) tubular reabsorption takes place in the proximal convoluted tubule.
19. Water is reabsorbed by osmosis.
20. Glucose is freely filtered from the blood into the filtrate, but reabsorption back into the blood is limited by the carrier molecules available for transport. When glucose levels exceed the renal threshold for reabsorption, the excess molecules remain in the urine.
21. Water is not reabsorbed in the ascending limb of the nephron loop.
22. Tubular secretion adds substances to the filtrate.
23. If blood volume and/or pressure decreases, urine volume is likely to decrease in an attempt to conserve fluids to return blood volume and/or pressure back to normal.

This compensation is accomplished by hormone action.

24. Three hormones that act on the kidney tubules are aldosterone, antidiuretic hormone, and atrial natriuretic hormone.

25. Renin is an enzyme that promotes the production of angiotensin II in the blood, which is a vasoconstrictor and has an effect on blood pressure.

26. Angiotensin II stimulates the cortex of the adrenal gland to secrete aldosterone.

27. Micturition is the act of expelling urine from the urinary bladder. It is also called urination or voiding.

28. Specific gravity of urine increases as the solute concentration increases.

29. a) Hematuria
    b) Glucosuria
    c) Albuminuria
    d) Pyuria

30. The intracellular fluid compartment has the most volume, accounting for 2/3 of the body fluid.

31. Most fluid intake is in beverages; most of the fluid loss is through the urine.

32. a) Na^+ is the predominant cation in ECF
    b) K^+ is the predominant cation in ICF
    c) Cl^- is the predominant anion in ECF
    d) Phosphates are ICF anions

33. A blood pH of 7.33 is indicative of acidosis.

34. If blood pH is too acid, the urinary system responds by secreting hydrogen ions from the kidney tubules. This makes the urine more acidic and removes the hydrogen ions from the body.

☞ **Answers to Learning Exercises**

Functions of the Urinary System (Objective 1)

1. Rids the body of wastes
   Regulates fluid volume
   Maintains electrolyte concentration
   Controls blood pH
   Secretes erythropoietin
   Secretes renin

Components of the Urinary System (Objectives 2-15)

1. Kidney, ureter, bladder, urethra.
2. D   Major calyx
   B   Minor calyx
   A   Renal capsule
   I   Renal column
   H   Renal cortex
   F   Renal papilla
   C   Renal pelvis
   G   Renal pyramid
   E   Ureter
3. B   Afferent arteriole
   E   Ascending limb
   D   Collecting duct
   J   Descending limb
   C   Distal convoluted tubule
   A   Efferent arteriole
   H   Glomerular capsule
   I   Glomerulus
   F   Nephron loop
   G   Proximal convoluted tubule
4. Macula densa
   Juxtaglomerular cells
   Renin
   Macula densa
   1200 ml/minute
   Segmental arteries
   Interlobar artery
   Arcuate artery
   Interlobular artery
   Afferent arteriole
   Peritubular capillaries
   Renal column
   Segmental veins
5. Ureters
   Transitional
   Urinary bladder
   Rugae
   Transitional
   Detrusor
   Trigone
   Smooth (involuntary)
   Skeletal (voluntary)
   Urethra
   Prostatic urethra
   Spongy (penile) urethra

Urine Formation (Objectives 16-28)

1. C   (Tubular cells to filtrate)
   A   (Glomerulus to capsule)
   B   (Tubules to capillaries)
2. Blue circles around GOP and CHP.
   Red circle around GHP.
   Net filtration pressure = 10 mm Hg.
   (GHP − GOP − CHP)
3. Aldosterone
   Antidiuretic hormone
   Atrial natriuretic hormone (atriopeptin)
   Renin
   Angiotensin
   Micturition
4. 1) Decreases or reduces
   2) Osmosis
   3) Active transport
   4) Carrier molecules

5) Threshold
6) Urine (filtrate)
7) Secretion
8) Hydrogen

5. It is powerful vasoconstrictor.
It increases aldosterone, which increases sodium and water reabsorption to increase blood volume, which increases blood pressure.

Characteristics of Urine (Objectives 29-31)

1. Checks (✓) before the following:
The color of urine is due to urochrome.
Urine is usually slightly acidic,...
High protein diets tend to make...
The specific gravity of urine...
The predominant solute in urine is urea.

2. Pyuria   (White blood cells)
Glucosuria   (Glucose)
Albuminuria   (Albumin)
Hematuria   (Erythrocytes)

Body Fluids (Objectives 32-40)

1. Beverages (* most)
Food
Metabolic water

2. Urine (* most)
Lungs
Skin
GI tract

3. D 20%   (Extracellular)
E 15-16%   (Interstitial)
C 40%   (Intracellular)
F 4-5%   (Plasma)
A 40%   (Solutes)
B 60%   (Total fluids)

4. Sodium
Chloride
Potassium
Phosphates
Aldosterone

5. 7.35-7.45
Alkalosis
Acidosis
Buffers
Respiratory system
Kidneys

☞ **Answers to Chapter Self-Quiz**

1. d
2. b
3. c
4. a
5. Detrusor
Trigone
External urethral sphincter
Prostatic urethra

6. Glomerular filtration: Capillary to tubule
Tubular reabsorption: Tubule to capillary
Tubular secretion: Capillary to tubule

7. Increase
Decrease
Increase
Decrease
Decrease

8. .6 x 80 = 46 liters of water
2/3 of 48 liters = 32 liters of ICF
or .4 x 80 = 32 liters of ICF
ECF = Total water − ICF = 16 liters
Interstitial fluid = 3/4 ECF = 12 liters

9. a
10. c

☞ **Answers to Terminology Exercises**

Nocturia
Cystopexy
Nephrolithiasis
Micturition
Juxtaglomerular

Suture the urinary bladder
Blood in the urine
Unable to hold urine
Lack of an opening in the urethra
Obstruction of the urethra

D (inflammation around the ureter)
C (glandular tumor of the kidney)
E (surgical removal of stone from pelvis)
B (painful urination)
A (dilation of the urinary bladder)

☞ **Answers to Fun and Games**

Afferent arteriole
Oliguria
ML
TOTAL: Glomerular filtration

Proximal
Plasma
Juxta
Atresia
Rugae
U
TOTAL: Juxtaglomerular apparatus

Cortex
Efferent arteriole
Medulla
Hilum
TOTAL: Extracellular fluid

☞ Quiz/Test Questions

Note: There are fifty multiple-choice questions for this chapter in the computerized test bank.

Name the following:

1. The striated appearing structures that make up the medulla of the kidney.
   Answer: renal pyramids.

2. Central collecting structure in the kidney that is continuous with the ureter.
   Answer: renal pelvis.

3. Cluster of capillaries in the renal corpuscle.
   Answer: glomerulus.

4. Portion of renal tubule adjacent to the glomerular capsule.
   Answer: proximal convoluted tubule.

5. Structure that forms where the ascending limb touches the afferent arteriole.
   Answer: juxtaglomerular apparatus.

6. Arteries that are located in the renal columns.
   Answer: interlobar arteries.

7. Collective term for the muscle in the wall of the urinary bladder.
   Answer: detrusor muscle.

8. Tubular structure that carries urine from the urinary bladder to the exterior.
   Answer: urethra.

9. Process by which plasma components leave the blood and enter the renal tubules.
   Answer: glomerular filtration.

10. Specific region where most of the tubular reabsorption occurs.
    Answer: proximal convoluted tubule.

11. Two hormones that have a direct influence urine concentration and volume.
    Answer: (any two) antidiuretic hormone (ADH), aldosterone, atrial natriuretic hormone.

12. Enzyme produced by the kidneys that promotes the production of angiotensin II.
    Answer: renin.

13. Another term for urination or voiding.
    Answer: micturition.

14. Presence of red blood cells in the urine.
    Answer: hematuria.

15. Fluid compartment that contains 2/3 of the total body fluid volume.
    Answer: intracellular fluid compartment.

True/False Questions:

1. Glucose normally passes freely from the blood into the glomerular capsule.
   Answer: True.

2. Urine volume increases as a result of increased ADH.
   Answer: False; ADH causes fluid retention and decreased urine volume.

3. Water is reabsorbed by osmosis from all parts of the renal tubule.
   Answer: False; it is reabsorbed by osmosis, but the ascending limb is impermeable to water.

4. The predominant solutes in urine are glucose and sodium.
   Answer: False; the predominant solute is urea.

5. About 40 percent of the body weight is fluid.
   Answer: False; 60 percent of body weight is fluid and 40 percent solutes.

6. Blood plasma accounts for most of the extracellular fluid volume.
   Answer: False; only 1/4 of ECF is plasma, 3/4 of ECF is interstitial fluid.

7. Potassium and phosphates are the major intracellular ions.
   Answer: True.

8. The respiratory system compensates for deviations in pH by adjusting the amount of carbon dioxide that is exhaled.
   Answer: True.

9. Buffers adjust deviations in pH by removing hydrogen ions from the body.
   Answer: False; urinary system removes hydrogen ions by tubular secretion.

10. An increase in pH above normal is termed acidosis.
    Answer: False; it is alkalosis.

19 Reproductive System

☞ **Key Terms/Concepts**

Corpora cavernosa Two dorsal columns of erectile tissue found in the penis.

Corpus luteum Under the influence of luteinizing hormone, the structure that develops from the mature follicle after ovulation.

Corpus spongiosum Ventral column of erectile tissue found in the penis.

Estrogen Hormone that is secreted by the ovarian follicles and is responsible for the development and maintenance of female secondary sex characteristics and the repair of the uterine lining after menstruation.

Gametes Sex cells; sperm and ova.

Gonads Primary reproductive organs; organs that produce the gametes; testes in the male and ovaries in the female.

Mammary glands Organs of milk production located within the breast.

Menarche First period of menstrual bleeding at puberty.

Menopause Cessation of menstrual bleeding; termination of uterine cycles.

Oogenesis Process of meiosis in the female in which one ovum and three polar bodies are produced from one primary oocyte.

Oogonia Stem cells that give rise to ova or egg cells.

Ovarian cycle Monthly cycle of events that occur in the ovary from puberty to menopause; occurs concurrently with the uterine cycle.

Ovarian follicle An oocyte surrounded by one or more layers of cells within the ovaries.

Polar body A small cell resulting from the unequal division of cytoplasm during the meiotic division of an oocyte.

Progesterone Hormone secreted by the corpus luteum that prepares the uterine lining for implantation of a developing embryo.

Puberty Period during which secondary sex characteristics begin to appear and capability for sexual reproduction becomes possible.

Pudendum Collective term for the external accessory structures of the female reproductive system; also called the vulva.

Seminiferous tubules Tightly coiled structures within which sperm are produced in the testes.

Spermatogenesis Process of meiosis in the male in which four spermatids are produced from one primary spermatocyte.

Spermatogonia Stem cells that give rise to sperm cells.

Spermiogenesis Morphological changes that transform a spermatid into a mature sperm.

Testosterone Principle male sex hormone that is responsible for the development and maintenance of male secondary sex characteristics.

Uterine cycle Monthly cycle of events that occur in the uterus from puberty to menopause; also called the menstrual cycle; occurs concurrently with the ovarian cycle.

Zygote The single diploid cell that is a fertilized ovum.

☞ **Chapter Objectives**

Upon completion of this chapter the student should be able to:

1. Distinguish between primary and secondary reproductive organs.
2. Label the parts of the male reproductive system on a diagram of a midsagittal section through the pelvis.
3. Describe the structure and location of the testes.
4. State the significance of seminiferous tubules and interstitial cells.
5. Draw and label a diagram or flow chart that illustrates spermatogenesis.
6. Name the process by which spermatids become spermatozoa and describe the three regions of a mature sperm.
7. Trace the pathway of sperm from the testes to the outside of the body.
8. Name three accessory glands of the male reproductive system and describe the contribution each makes to the seminal fluid.
9. Distinguish between the emission and ejaculation of semen.
10. Use the terms corpora cavernosa, corpus spongiosum, root, body, and glans penis to describe the structure of the penis.

11. Outline the physiological events in the male sexual response.

12. Describe the role of GnRH, FSH, LH, and testosterone in the male reproductive system.

13. Label the parts of the female reproductive system on a diagram of a midsagittal section through the pelvis.

14. Describe the location and structure of the ovaries.

15. Draw and label a diagram or flow chart that illustrates oogenesis.

16. Compare oogenesis and spermatogenesis on the basis of when the different stages occur and the final results.

17. Describe the development of ovarian follicles as they progress from primordial follicles to primary follicles, secondary follicles, vesicular follicles, corpus luteum, and finally, the corpus albicans.

18. Identify the stage of oogenesis that exists at the time of ovulation.

19. Use the terms infundibulum and fimbriae in a description of the uterine tubes.

20. Use the terms fundus, body, cervix, internal os, external os, broad ligament, and anteflexed to describe features of the uterus.

21. Name the three layers of the uterine wall and describe the tissue in each one; distinguish between the two parts of the endometrium.

22. State three functions of the vagina.

23. Name six structures included in the vulva.

24. Name two accessory glands of the female reproductive system that are associated with the vestibule.

25. Outline the physiological events in the female sexual response.

26. Describe the roles of GnRH, FSH, LH, estrogen, and progesterone in the female reproductive system.

27. Define the terms menarche and menopause.

28. Name the three phases of the ovarian cycle, state when each phase occurs, and describe what happens in each phase.

29. Name the three phases of the uterine cycle, state when each phase occurs, and describe what happens in each phase.

30. Explain how the events of the ovarian cycle affect the uterine cycle.

31. Describe the hormonal changes that occur during menopause.

32. Describe the location and structure of the mammary glands.

33. Describe the effects of estrogen, progesterone, prolactin, and oxytocin on the mammary glands.

☞ Chapter Outline/Summary

Male Reproductive System (Objectives 1-12)
Testes (Objectives 1-6)
- Structure (Objectives 1-4)
  - Testes are the male gonads. They begin development in the abdominal cavity, then descend into the scrotum shortly before birth.
  - The testes are surrounded by the tunica albuginea, which extends inward to divide the organ into lobules.
  - Each lobule contains seminiferous tubules and interstitial cells.
  - Sperm are produced in the seminiferous tubules and the interstitial cells produce testosterone.
- Spermatogenesis (Objectives 5 and 6)
  - Spermatogenesis, which begins at puberty, is the process by which spermatids are formed. Spermatogonia become primary spermatocytes and each spermatocyte produces four spermatids by meiosis.
  - Spermiogenesis changes spermatids into mature spermatozoa, each with a head, midpiece, and tail. The head contains the nucleus, the midpiece contains mitochondria, and the tail is a flagellum for movement.

Duct system (Objective 7)
- The epididymis is a convoluted tube on the margin of the testis. Sperm mature and become fertile in the epididymis.
- The ductus deferens begins at the epididymis and extends to the ejaculatory duct posterior to the urinary bladder.
- The ejaculatory duct is a short passageway that is formed when the ductus deferens and duct from the seminal vesicles join. It penetrates the prostate gland and empties into the urethra.
- The male urethra is a passageway for sperm, fluids from the reproductive system, and urine. It is divided into the prostatic urethra, membranous urethra, and penile urethra.

Accessory glands (Objectives 8 and 9)
- Seminal vesicles (Objective 8)
  - Seminal vesicles are paired, saccular glands posterior to the urinary bladder.
  - Secretion contains fructose, prostaglandins, and coagulation proteins.
  - Secretion from these glands contributes over 60% of the volume of seminal fluid.

- Prostate (Objective 8)
  - The prostate is a firm dense gland located inferior to the urinary bladder; it encircles the proximal part of the urethra.
  - Secretions are alkaline and enhance motility of sperm.
- Bulbourethral glands (Objective 8)
  - The bulbourethral glands are located near the base of the penis and secrete a viscous alkaline fluid.
  - Secretion neutralizes acidity of vagina and provides lubrication during intercourse.
- Seminal fluid (Objective 9)
  - Seminal fluid or semen is a mixture of sperm cells and the secretions of the accessory glands.
  - Emission is the discharge of semen into the urethra; ejaculation is the forceful expulsion of semen from the urethra.

Penis (Objective 10)
- The penis consists of three columns of erectile tissue. The two dorsal columns are the corpora cavernosa and the ventral column is corpus spongiosum.
- The root of the penis attaches it to the pubic arch, the body is the visible portion, and the glans penis is the expanded tip of corpus spongiosum.
- The urethra extends through the entire length of the penis.

Male Sexual Response (Objective 11)
- The male sexual response includes erection and orgasm accompanied by ejaculation of semen.
- Orgasm is followed by a variable time period during which it is not possible to achieve another erection.

Hormonal control (Objective 12)
- At puberty, the hypothalamus secretes GnRH which stimulates the anterior pituitary to secrete FSH and LH.
- FSH stimulates the seminiferous tubules and spermatogenesis.
- LH stimulates the interstitial cells and the production of testosterone.
- Testosterone from the interstitial cells stimulates the development of the secondary sex characteristics and spermatogenesis.

Female Reproductive System (Objectives 13-33)

Ovaries (Objectives 13-18)
- Structure (Objective 14)
  - The female gonads are the ovaries, which are located on each side of the uterus in the pelvic cavity.
  - The ovaries are covered with simple cuboidal epithelium around the tunica albuginea.
  - Numerous ovarian follicles make the cortex appear granular; the medulla is connective tissue with vessels and nerves.
- Oogenesis (Objectives 15 and 16)
  - Oogenesis begins in prenatal development with the formation of the primary oocyte. Division ceases in this stage and the oocytes remain dormant until puberty.
  - Beginning at puberty, each month a primary oocyte resumes meiosis and produces a secondary oocyte and a polar body. Division again halts.
  - If a sperm penetrates the oocyte, meiosis resumes and a mature egg and another polar body are produced.
  - Oogenesis differs from spermatogenesis in that each primary oocyte produces one ovum and 3 nonfunctional polar bodies; each primary spermatocyte produces four spermatids. Beginning at puberty, spermatogenesis is a continuous process; oogenesis occurs in monthly cycles. Spermatogenesis doesn't begin until puberty; oogenesis begins in prenatal development.
- Ovarian follicle development (Objective 17)
  - An ovarian follicle consists of an oocyte with one or more layers of cells surrounding it.
  - Primordial follicles, each with a primary oocyte surrounded by a single layer of cells, are the follicles present at birth.
  - At puberty the primordial follicles begin to grow, become secondary follicles, and some mature and become vesicular follicles.
  - Vesicular follicles rupture at ovulation and release their secondary oocyte.
  - Follicles develop under the influence of FSH; follicle cells produce estrogen.
  - After ovulation, the follicle cells are transformed into a corpus luteum that produces progesterone. The corpus luteum degenerates into a corpus albicans.
- Ovulation (Objective 18)
  - At ovulation, a secondary oocyte is discharged from a vesicular follicle.
  - The ovulated oocyte is surrounded by a noncellular zona pellucida and several layers of cells called the corona radiata.

Genital tract (Objectives 19-22)

- Uterine tubes (Objective 19)
  - The uterine tubes, also called fallopian tubes or oviducts, extend laterally from each side of the uterus.
  - The end of the uterine tube near the ovary expands to form the infundibulum. Fingerlike projections, called fimbriae, extend from the infundibulum.
- Uterus (Objectives 20 and 21)
  - The uterus consists of a fundus, body, and cervix. The broad ligament is a fold of peritoneum that extends laterally from the uterus to the pelvic wall.
  - The fundus and body of the uterus are normally anteflexed over the superior surface of the urinary bladder.
  - The internal os is the opening from the body into the cervix; the external os is the opening from the cervix into the vagina.
  - Visceral peritoneum forms the perimetrium, the outer layer of the uterine wall; smooth muscle makes up the thick myometrium; and the endometrium is mucous membrane. The endometrium is separated into a deeper stratum basale and superficial stratum functionale.
- Vagina (Objective 22)
  - The vagina extends from the cervix to the exterior.
  - The vagina serves as a passageway for menstrual flow, receives the erect penis during intercourse, and is the birth canal during the birth of a baby.

External genitalia (Objectives 23 and 24)

- Collectively, the female external genitalia are referred to as the vulva or pudendum.
- The external genitalia include the labia majora, mons pubis, labia minora, clitoris, and accessory glands.
- The area between the two labia minora is the vestibule. The clitoris (erectile tissue) is at the anterior end of the vestibule.
- Paraurethral and greater vestibular glands, accessory glands of the female reproductive tract, open into the vestibule. The urethra and vagina also open into the vestibule.

Female sexual response (Objective 25)

- The female sexual response includes erection and orgasm, but there is no ejaculation.
- A woman may become pregnant without having an orgasm.

Hormonal control (Objectives 26-30)

- GnRH, FSH, LH, estrogen, and progesterone interact to create the ovarian and uterine cycles.
- Ovarian cycle
  - The monthly ovarian cycle begins with the follicle development during the follicular phase, continues with ovulation during the ovulatory phase, and concludes with the development and regression of the corpus luteum during the luteal phase.
  - The follicle develops under the influence of FSH. As the follicle matures, it secretes increasing amounts of estrogen. At ovulation, the estrogen level falls, then the corpus luteum, under the influence of LH, begins secreting progesterone.
- Uterine (menstrual) cycle
  - Menarche is the first menstrual flow; menopause is the time when the monthly ovarian and uterine cycles cease.
  - The uterine cycle takes place simultaneously with the ovarian cycle.
  - The uterine cycle begins with menstruation during the menstrual phase, continues with repair of the endometrium during the proliferative phase, and ends with the growth of glands and blood vessels during the secretory phase.
  - The menstrual phase is the result of decreased amounts of progesterone and estrogen from the corpus luteum; estrogen from developing follicles is responsible for the proliferative phase; and progesterone from the corpus luteum is responsible for the secretory phase.

Mammary glands (Objectives 32 and 33)

- The mammary glands, located with the breast, consist of lobules of glandular units that produce milk. Lactiferous ducts transport the milk to the nipple.
- Cords of connective tissue, called suspensory ligaments, help support the breast.
- Estrogen and progesterone stimulate the development of glandular tissue and ducts in the breast. Prolactin stimulates the production of milk. Oxytocin causes the ejection of the milk from the breast.

☞ **Answers to Review Questions**

1. The primary reproductive organs, also called gonads, are the testes and ovaries. These organs are responsible for producing the gametes, the sperm and egg cells, and for producing hormones.

2. See Figure 19-1 in the textbook.

3. The male gonads begin development in the high in abdominal cavity near the kidneys. Shortly before birth they descend into the scrotum.

4. The seminiferous tubules are highly coiled tubules within the testes. Sperm are produced within the seminiferous tubules.

5. a) 46 (23 pr) chromosomes.
   b) 46 (23 pr) chromosomes.
   c) 23 chromosomes.
   d) 23 chromosomes.
   e) 23 chromosomes.

6. a) Acrosome contains enzymes (hyaluronidase) that help sperm penetrate female gamete.
   b) Head is the nuclear region and contains the chromosomes.
   c) Midpiece is the metabolic region and contains the mitochondria to provide ATP for the sperm.

7. a) Ductus deferens
   b) Prostatic urethra

8. Seminal vesicles contribute about 60% of the volume of the seminal fluid. These glands are posterior to the urinary bladder.

9. Emission is the forceful discharge of seminal fluid into the urethra.

10. Corpus spongiosum surrounds the urethra in the penis.

11. Parasympathetic reflexes dilate the arterioles that supply blood to the erectile tissue and constrict the veins that remove the blood. This results in an erection. As the parasympathetic impulses continue, they prompt a surge of sympathetic impulses to the genital organs, which results in emission of seminal fluid and this is immediately followed by ejaculation.

12. FSH works with testosterone to stimulate spermatogenesis in the seminiferous tubules. LH promotes the growth of the interstitial cells and stimulates these cells to produce testosterone.

13. See Figure 19-7 in the textbook.

14. Primordial follicles are present in the ovary at birth.

15. One secondary oocyte is produced from each primary oocyte, the other cell that is produced is a nonfunctional polar body. In the male two secondary spermatocytes are produced from each primary spermatocyte.

16. A noncellular zona pellucida and a cellular corona radiata surround the secondary oocyte in a vesicular follicle.

17. A secondary oocyte in metaphase of the second meiotic division is released at the time of ovulation.

18. The distal portion, or infundibulum, is the portion of the uterine tube closest to the ovary.

19. The cervix of the uterus is between the internal os and the external os.

20. The stratum functionale of the endometrium is the portion of the uterine wall that is shed during menstruation.

21. The vagina serves as a passageway for menstrual flow, receives the erect penis during intercourse, and is the birth canal during the birth of a baby. It opens to the outside through the vaginal orifice in the region of the vestibule.

22. The greater vestibular glands are adjacent to the vaginal orifice. They produce a mucus-like secretion for lubrication during intercourse.

23. Parasympathetic impulses cause an erection of the tissue in the clitoris, vaginal mucosa, breasts, and nipples. Sympathetic responses produce contractions of the uterus and muscles of the pelvic floor.

24. Estrogen is increasing during the follicular phase of the ovarian cycle.

25. The secretory phase of the uterine cycle corresponds to the luteal phase of the ovarian cycle. Progesterone is increasing during this time and there is some increase in estrogen from the corpus luteum.

26. The monthly menstrual cycles cease at menopause because the ovarian follicles stop responding to FSH and LH from the pituitary gland. This results in a reduction of estrogen and progesterone so the cyclic changes do not occur in the uterus.

27. Estrogen stimulates the development of glandular tissue and the deposition of adipose in breast. Progesterone stimulates the development of the duct system.

28. Oxytocin from the posterior pituitary gland causes the ejection of milk from the mammary gland.

Answers to Learning Exercises

Male Reproductive System (Objectives 1-12)

1. D (Bulbourethral gland)
   M (Corpus cavernosum)
   E (Corpus spongiosum)
   F (Ductus deferens)
   C (Ejaculatory duct)
   L (Epididymis)
   N (Glans penis)
   J (Prostate)
   H (Scrotum)
   I (Symphysis pubis)
   G (Testicle)
   B (Urethra)
   K (Urethra)
   A (Urinary bladder)
2. 6 (Ductus deferens)
   4 (Efferent ducts)
   7 (Ejaculatory duct)
   5 (Epididymis)
   9 (Membranous urethra)
   10 (Penile urethra)
   8 (Prostatic urethra)
   3 (Rete testis
   1 (Seminiferous tubules)
   2 (Straight tubules)
3. Testes
   Spermatozoa (sperm)
   Scrotum
   Dartos
   Cremaster
   Tunica albuginea
   Seminiferous tubules
   Interstitial cells
   Spermiogenesis
   Midpiece
   23
   Supporting cells (Sertoli)
4. C (Product secreted...)
   B (Encircles...)
   A (Located posterior...)
   A (Secretion accounts...)
   C (Located near...)
   C (Smallest of...)
   A (Secretion has...)
5. 1) Parasympathetic
   2) Arteries
   3) Veins
   4) Erection
   5) Sympathetic
   6) Emission
   7) Ejaculation
   8) Increased
   9) Increased
   10) Increased respiration
   11) Orgasm or climax

6. GnRH
   Testosterone
   LH or ICSH
   Androgens
   FSH
   LH
   FSH
   Adrenal cortex
   Spermatogenesis
   Development and maintenance of secondary sex characteristics

Female Reproductive System (Objectives 13-33)

1. I (Cervix)
   G (Clitoris)
   H (Infundibulum)
   F (Mons pubis)
   A (Ovary)
   J (Rectum)
   E (Symphysis pubis)
   L (Urethra)
   D (Urinary bladder)
   B (Uterine tube)
   C (Uterus
   K (Vagina)
2. Ovary
   Primary oocyte
   Secondary oocyte
   Zona pellucida
   Corona radiata
   Corpus luteum
   Fimbriae
   External os
   Broad ligament
   Myometrium
   Stratum functionale
   Hymen
3. D (Antrum
   J (Corpus albicans)
   I (Corpus luteum)
   G (Fimbriae)
   A (Infundibulum)
   F (Primary follicle)
   E (Secondary follicle)
   B (Secondary oocyte)
   H (Uterine tube)
   C (Vesicular follicle)
4. Vulva or pudendum
   Labia majora
   Mons pubis
   Vestibule
   Clitoris
   Paraurethral glands
   Greater vestibular glands

5. True   (The female...)
   False   (Sympathetic responses...)
   True   (Sympathetic responses...)
6. A   (Starts the events...)
   B   (Stimulates growth...)
   A   (Secreted by hypothalamus)
   B, C   (Secreted by anterior pituitary)
   D   (Secreted by cells...)
   D, E   (Secreted by corpus luteum)
   C   (Triggers ovulation)
   E   (Stimulates secretory phase...)
   D   (Stimulates proliferative...)
   C   (Stimulates development...)
   D   (Stimulates development...)
   E   (Stimulates development...)
   D   (Causes an accumulation...)
   B, C   (Levels increase...)
   D, E   (Levels decrease...)
7. Increases
   Increases
   Increases
   Decreases
   Decreases
8. Areola
   Suspensory (Cooper's) ligaments
   Lactiferous duct
   Lactiferous sinus (ampulla)
   Prolactin
   Oxytocin

☞ Answers to Chapter Self-Quiz

1. b
2. d
3. P (Encircles...)
   S (Contributes...)
   B (Empties...)
   P (Secretion...)
   S (Empties...)
4. Should have X before the following:
   Erectile tissue
   Encircles the urethra
   Makes up the glans penis
5. C (Stimulates...)
   B (Hormone...)
   D (Secreted...)
   C (Stimulates...)
   E (Promotes...)
   B (Stimulates...)
   D (Promotes...)
   A (Stimulates...)
   F (Promotes...)
   A (With testosterone...)
6. b
7. d
8. Just prior to ovulation.
9. d

10. **Estrogen**: deposition of adipose and development of glandular material.
    Progesterone: development of the duct system.
    Prolactin: milk production
    Oxytocin: milk ejection

☞ Answers to Terminology Exercises

Gynecology
Balanitis
Androgenic
Myometrium
Mammoplasty

Surgical excision of the uterus
Inflammation of the uterine tube
Enlarged prostate gland
Surgical excision of the ovary
Without (lacking) a testicle

D (urethra opens on dorsum of penis)
A (surgical repair of the vagina and perineum)
E (excessive monthly flow)
B (suture of the vagina)
C (difficult or painful monthly flow)

☞ Answers to Fun and Games

N   True
N   False
U   False
R   True
D   True
C   True
O   False
I   False
O   True
Y   False
E   True
T   False
O   True
N   False
T   True
P   True
I   True
T   False
U   True
C   False
I   False
R   True

True letters: N R D C O E O T P I U R
Word: Reproduction
False letters: N U O I Y T N T C T
Word: Continuity

☞ Quiz/Test Questions

Note: There are fifty multiple-choice questions for this chapter in the computerized test bank.

Name the following:

1. The specific structures where sperm are produced.
   Answer: seminiferous tubules of the testes.

2. Hormone that is produced by the interstitial cells of the testes.
   Answer: testosterone.

3. Structure along the margin of the testis where sperm mature and are stored.
   Answer: epididymis.

4. Tubular structure within the spermatic cord and abdominopelvic cavity that conveys sperm.
   Answer: ductus deferens.

5. Accessory gland that contributes the greatest volume to the seminal fluid.
   Answer: seminal vesicles.

6. Portion of the urethra that receives fluid from the ejaculatory duct.
   Answer: prostatic urethra.

7. Type of erectile tissue that surrounds the urethra.
   Answer: corpus spongiosum.

8. Stage of oogenesis that is ovulated.
   Answer: secondary oocyte.

9. Expanded distal region of the uterine tube.
   Answer: infundibulum.

10. Muscular layer of the uterine wall.
    Answer: myometrium.

11. Portion of the uterine wall that sloughs off during menstruation.
    Answer: stratum functionale of the endometrium.

12. Region of the uterus that projects into the vagina.
    Answer: cervix.

13. Hormone that triggers ovulation.
    Answer: luteinizing hormone.

14. First menstrual flow.
    Answer: menarche.

15. Phase of the ovarian cycle that corresponds to the proliferative phase of the uterine cycle.
    Answer: follicular phase.

True/False Questions:

1. Seminal fluid is an alkaline mixture of sperm and secretions from the accessory glands.
   Answer: True.

2. The distal end of corpus cavernosum is expanded to form the glans penis.
   Answer: False; it is corpus spongiosum.

3. An erection in the male sexual response is due to sympathetic impulses.
   Answer: False; it is due to parasympathetic impulses.

4. Emission and ejaculation are impossible in a male who has had a vasectomy because the pathway is blocked.
   Answer: False; emission and ejaculation of seminal fluid are possible, however there are no sperm present.

5. Testosterone is responsible for the development and maintenance of male secondary sex characteristics.
   Answer: True.

6. In female cycles, more progesterone is produced in the two weeks before ovulation than is produced in the two weeks after ovulation.
   Answer: False; progesterone is produced by the corpus luteum which develops after ovulation.

7. The uterine lining is thicker on day 21 of the ovarian cycle than it is on day 7 of the ovarian cycle.
   Answer: True.

8. Both sympathetic and parasympathetic nerve impulses are involved in the female sexual response.
   Answer: True.

9. Prolactin from the anterior pituitary gland promotes the production and release of milk from the mammary gland.
   Answer: False; prolactin promotes the production of milk, oxytocin is necessary for ejection.

10. Levels of FSH decline after menopause.
    Answer: False; FSH increases because of a lack of ovarian hormone feedback.

20 Development

Key Terms/Concepts

Amnion The innermost fetal membrane; transparent sac that holds the developing fetus suspended in fluid.

Blastocyst Hollow sphere of cells that forms when a cavity develops in a morula, usually present by the fifth day after conception.

Capacitation Process that enables sperm to penetrate an egg; membrane around the acrosome weakens so the enzymes can be released.

Chorion Outermost extraembryonic, or fetal, membrane; contributes to the formation of the placenta.

Cleavage Series of mitotic cell divisions after fertilization; resulting cells are called blastomeres.

Embryo Stage of development that lasts from the beginning of the third week to the end of the eighth week after fertilization; period during which the organ systems develop in the body.

Embryonic disk Cells of the early embryo that give rise to the three primary germ layers.

Fetus Term used for the developing offspring from the beginning of the 9th week after fertilization until birth.

Human chorionic gonadotropin (HCG) Hormone secreted by the trophoblast, which has an effect similar to LH and causes the corpus luteum to remain functional to maintain pregnancy.

Implantation Process by which the developing embryo becomes embedded in the uterine wall; usually takes about a week and is completed by the 14th day after fertilization.

Parturition Act of giving birth to an infant.

Placenta Structure that anchors the developing fetus to the uterus and provides for the exchange of gases, nutrients, and waste products between the maternal and fetal circulations.

Postnatal development Development which begins with birth and lasts until death.

Prenatal development Development within the uterus.

Senescence Period of old age.

Zygote Diploid cell that is the result of fertilization; fertilized egg.

Chapter Objectives

Upon completion of this chapter the student should be able to:

1. Define prenatal development and postnatal development.
2. Define the term capacitation.
3. Describe the events in the process of fertilization, state where it normally occurs, and name of the cell that is formed as a result of fertilization.
4. Name the three divisions of prenatal development and state the period of time for each.
5. Describe three significant developments that take place during the preembryonic period.
6. Define the terms cleavage, blastomeres, morula, blastocyst, blastocoele, trophoblast, and inner cell mass.
7. Describe where implantation normally occurs and how it saves itself from being aborted.
8. Name the three primary germ layers.
9. Describe three significant developments that take place during the embryonic period.
10. Name the four extraembryonic membranes and describe the function of each one.
11. Describe the formation, structure, and functions of the placenta.
12. List five derivatives from each of the primary germ layers.
13. State the two fundamental processes that take place during fetal development.
14. Name, describe the location, and state the function of five structures that are unique in the circulatory pattern of the fetus.
15. Define the terms gestation, parturition, and labor.
16. State the length of the normal gestation period from the time of the last menstrual period and from the time of fertilization.
17. Describe the roles of the hypothalamus, estrogen, progesterone, oxytocin, and prostaglandins in promoting labor.
18. Describe the three stages of labor.
19. Describe the changes that take place in the baby's respiratory system and circulatory pathway at birth or soon after birth.
20. Distinguish between colostrum and milk.
21. Explain the relationship between a baby's suckling, the mother's hypothalamus, and milk production and ejection.

22. Name and define six periods in postnatal development.

☞ Chapter Outline/Summary

Fertilization (Objectives 1-4)

- Prenatal development is the period from fertilization to birth. Postnatal development is the period from birth to death.
- As the sperm move through the female reproductive tract, the acrosome membrane weakens. This is called capacitation.
- The process of fertilization begins when a single sperm penetrates the cell membrane of a secondary oocyte. This stimulates completion of the second meiotic division which produces a second polar body and an ovum. The nuclear membranes of the male and female pronuclei degenerate, the two nuclei fuse.
- The fertilized egg, which has a full complement of 46 chromosomes, is called a zygote.
- Fertilization normally takes place in the uterine tube.
- Prenatal development consists of a pre-embryonic period (2 weeks), embryonic period (6 weeks), and fetal period (30 weeks).

Preembryonic period (Two weeks)
(Objectives 5-8)
Cleavage (Objective 6)

- Cleavage is a rapid series of mitotic cell divisions after fertilization. The cells that result are blastomeres.
- A solid ball of blastomeres is a morula. A cavity forms inside the morula and it becomes a blastocyst.
- The cavity inside the blastocyst is the blastocoele and the cells around the outside are the trophoblast. A cluster of cells on one side is the inner cell mass and represents the future embryo.

Implantation (Objective 7)

- Implantation occurs as endometrial tissue grows around the blastocyst. The entire process takes about seven days.
- The trophoblast cells secrete human chorionic gonadotropin, which acts like LH to maintain the corpus luteum. Since the corpus luteum continues to secrete progesterone, the uterine lining is maintained and menstruation is inhibited.

Formation of primary germ layers (Objective 8)

- The thee primary germ layers develop while implantation is taking place.
- The primary germ layers are ectoderm, mesoderm, and endoderm.

Embryonic development (Six weeks)
(Objectives 9-12)
Formation of the extraembryonic membranes (Objective 10)

- The amnion, chorion, yolk sac, and allantois are membranes that form outside the embryo and are called extraembryonic membranes.
- The extraembryonic membranes function in protection, nutrition, and excretion for the embryo.
- The amnion forms from the outer layer of the inner cell mass. It forms a fluid filled sac that surrounds the growing embryo. It helps maintain constant temperature and pressure around the embryo, provides for symmetrical development, and movement.
- The chorion develops from the trophoblast and contributes to the formation of the placenta.
- The yolk sac develops from the endoderm side of the embryonic disk and produces the primordial germ cells.
- The allantois develops from the yolk sac and contributes to the formation of the umbilical arteries and vein.

Formation of the placenta (Objective 11)

- The placenta develops from the endometrium of the uterus and the chorion of the embryo.
- Chorionic villi grow into the endometrium and blood filled lacunae from the endometrium surround the villi.
- Nutrients and oxygen diffuse from the mother's blood in the lacunae into the blood vessels in the chorionic villi. Waste materials diffuse in the opposite direction.
- The placenta functions as a temporary endocrine gland for the mother and produces estrogen and progesterone.

Organogenesis (Objective 12)

- All body organs develop from the ectoderm, mesoderm, and endoderm that are formed during the preembryonic period.
- All organ systems are formed by the end of the embryonic period.

Fetal Development (30 weeks)
(Objectives 13 and 14)

- The fetal period is one of growth and maturation of the organ systems that form during the embryonic period.
- Since the lungs and liver are nonfunctional during the fetal period, special structures in the circulatory pathway allow blood to bypass these organs. Other vessels take blood to and from the placenta for gaseous exchange.
- Two umbilical arteries carry fetal blood to the placenta and one umbilical vein returns

the oxygenated blood to fetal circulation. The ductus venosus carries blood from the umbilical vein to the IVC and bypasses the liver. The foramen ovale, in the interatrial septum, and the ductus arteriosus, between the pulmonary trunk and descending aorta, allow blood to bypass the lungs.

Parturition and Lactation (Objectives 15-21)
Labor and delivery (Objectives 15-18)

- Gestation period is the time from fertilization to birth; it is the time of pregnancy. It normally lasts for 266 days from fertilization for 280 days from the beginning of the last menstrual period.
- Parturition refers to the birth of a baby and labor is the process by which forceful contractions expel the fetus from the uterus.
- Near the end of gestation, estrogen levels increase and progesterone starts to decrease. This removes progesterone's inhibitory effects on the uterus. Estrogen also sensitizes oxytocin receptors. Pressure of the baby's head on the cervix signals the hypothalamus to secrete oxytocin and this, with prostaglandins, stimulates uterine contractions.
- The dilation stage of labor begins with the onset of true labor and lasts until the cervix is fully dilated. The expulsion stage lasts from full cervical dilation until delivery of the fetus. The final phase is the placental stage when the placenta and extraembryonic membranes are expelled.

Adjustments of the baby at birth (Objective 19)

- When the umbilical cord is cut, the baby's oxygen supply from the mother is terminated. Changes in blood gases stimulate the respiratory center. The first breath needs to be strong and deep to inflate the lungs.
- The special features of fetal circulation cease to function after birth and degenerate or change into their postnatal state at birth or soon after.

Physiology of lactation (Objectives 20 and 21)

- Lactation refers to the production of milk by the mammary glands.
- For the first 2 or 3 days the mammary glands secrete colostrum. After this, the glands produce milk.
- The baby's suckling at the nipple sends signals to the hypothalamus, oxytocin is released, and this ejects milk from the mammary glands.
- In response to baby's suckling, the hypothalamus also releases Prolactin releasing hor

mone. This creates a surge in prolactin, which stimulates milk production for the next feeding period.

Postnatal Development (Objective 22)

- Neonatal period begins at the moment of birth and lasts until the end of the first four weeks.
- Infancy lasts from the end of the first month to the end of the first year.
- Childhood lasts from the end of the first year until puberty.
- Adolescence begins at puberty and lasts until adulthood.
- Adulthood is the period from adolescence to old age.
- Senescence is the period of old age and ends in death.

☞ **Answers to Review Questions**

1. Prenatal development lasts for about 38 weeks and is the period before birth. The period from birth until death is the period of postnatal development.
2. The acrosomal enzymes break down the corona radiata and zona pellucida to create an opening for the sperm.
3. Fertilization usually takes place in the uterine tube on the 14th day of the menstrual cycle.
4. The single cell that results from fertilization is a zygote with 46 chromosomes.
5. The preembryonic period lasts for two weeks. Cleavage, implantation, and the formation of the primary germ layers occur during this time.
6. The three parts of a blastocyst are the blastocoele, trophoblast, and inner cell mass. The blastocoele is the cavity; the trophoblast is the cells around the periphery, and inner cell mass is a cluster of cells along one side.
7. The trophoblast secretes human chorionic gonadotropin. This hormone has an action similar to LH because it causes the corpus luteum to remain functional and secrete progesterone to maintain the endometrium.
8. Ectoderm:
   Epidermis of the skin
   Hair, nails, skin glands
   Lens of the eye
   Enamel of the teeth
   All nervous tissue
   Adrenal medulla
   Sense organ receptor cells
   Linings of the oral and nasal cavities, vagina, and anal canal

Mesoderm:
  Dermis of the skin
  Skeletal, smooth, cardiac muscle
  Connective tissue including cartilage and bone

  Epithelium of serous membranes
  Epithelium of joint cavities
  Epithelium of blood vessels
  Kidneys and ureters
  Adrenal cortex
  Epithelium of gonads and reproductive ducts
Endoderm:
  Epithelial lining of digestive tract
  Epithelium of the liver and pancreas
  Epithelium of urinary bladder and urethra
  Epithelium of the respiratory tract
  Thyroid, parathyroid, and thymus glands

9. The period of embryonic development lasts for six weeks. Significant developments include the formation of the extraembryonic membranes, formation of the placenta, and the formation of all organ systems in the body.

10. **Amnion:** The amnion is a fluid-filled sac that surrounds the developing embryo. The fluid cushions and protects the embryo, maintains a constant temperature and pressure around it, provides a medium for symmetric development, and allows freedom of movement.
Chorion: The chorion contributes to the formation of the placenta.
Yolk sac: The yolk sac produces the primordial germ cells, which migrate to the developing gonad, and produces blood until the liver is sufficiently developed.
Allantois: The allantois contributes to the development of the urinary bladder and the umbilical vessels.
Placenta: The placenta is site of gaseous exchange between the fetal blood and maternal blood.

11. The fetal period of development lasts about 30 weeks, from the beginning of the ninth week until parturition. It is period of growth and maturation of the organ systems.

12. a) Umbilical arteries carry fetal blood to the placenta.
b) The umbilical vein carries blood from the placenta back to the fetus.
c) The ductus venosus is a shunt that allows fetal blood to bypass the liver.
d) The foramen ovale allows blood to pass directly from the right atrium into the left atrium to bypass the lungs.
e) The ductus arteriosus allows blood to pass from the pulmonary trunk to the descending aorta to bypass the lungs.

13. The normal length of time from fertilization until parturition is 266 days.

14. Oxytocin from the posterior pituitary gland stimulates uterine contractions during labor.

15. **Dilation stage:** characterized by rhythmic and forceful uterine contractions, rupture of the amniotic sac, and dilation of the cervix.
Expulsion stage: lasts from full cervical dilation until delivery of the fetus.
Placental stage: placenta separates from the uterine wall and is expelled with the other membranes as the afterbirth.

16. A baby's first breath is normally strong and deep to inflate the lungs.

17. a) Umbilical arteries degenerate and become the lateral umbilical ligaments.
b) Umbilical vein degenerates and becomes the ligamentum teres of the liver.
c) Ductus venosus becomes the ligamentum venosum of the liver.
d) Foramen ovale closes and becomes the fossa ovalis in the interatrial septum.
e) Ductus arteriosus degenerates and becomes a fibrous ligamentum arteriosum.

18. Colostrum is produced for two or three days after birth, prior to milk production. It is a cloudy yellowish fluid that has less lactose than milk, almost no fat, and more protein, vitamin A and minerals than milk.

19. A baby's sucking action triggers nerve impulses that go to the hypothalamus, which signals the posterior pituitary to release oxytocin for the ejection of milk.

20. **Neonatal** begins at birth and lasts until the end of the first 4 weeks.
Infancy lasts from the end of the first month to the end of the first year.
Childhood lasts from the end of the first year until puberty.
Adolescence lasts from puberty until adulthood.
Adulthood is the period from adolescence to old age. It is a period of maintenance of body tissues, then a gradual degenerative change into senescence.
Senescence is the period of old age and ends in death. The body becomes less and less capable of coping with the demands placed on it.

☞ **Answers to Learning Exercises**

Fertilization (**Objectives 1-4**)
1. Zygote
   Corona radiata
   Capacitation
   24 hours
   Uterine tube

Preembryonic Period (Objectives 5-8)
1. Cleavage
   Implantation
   Formation of germ layers
   Cleavage
   Blastomeres
   Morula
   Blastocyst
   Inner cell mass
   Blastocoele
   Trophoblast
   Stratum functionale of endometrium
   Human chorionic gonadotropin (HCG)
   Progesterone
   Inner cell mass
   Ectoderm
   Mesoderm
   Endoderm

Embryonic Development (Objectives 9-12)
1. The period of embryonic development lasts from the beginning of the **third** week after conception to the end of the **eighth** week. Three significant developments during this period are the formation of the **extra-embryonic** membranes, formation of the **placenta**, and formation of all the body **organ** systems. During this period the developing offspring is called an **embryo**.
2. A (Forms a sac...)
   D (Produces the primordial...)
   B (Becomes part...)
   C (Develops from...)
   A (Cushions and protects...)
   C (Contributes to...)
   C (Develops fingerlike...)
   A (Filled with fluid)
3. 1) Chorionic villi
   2) Endometrium
   3) Chorionic villi
   4) Umbilical
   5) Blood
   6) Oxygen
   7) Carbon dioxide
   8) Fetal blood
   9) Maternal blood
4. C (Epithelial lining the digestive tract)
   A (Epidermis of the skin)
   B (Cardiac muscle)
   A (Nervous tissue)

B (Cartilage)
A (Hair, nails, glands of the skin)
C (Respiratory epithelium)
B (Epithelium lining the blood vessels)
A (Lining of the oral cavity)
B (Bone)
B (Dermis of the skin)
A (Epithelial lining of the vagina)

Fetal Development (Objectives 13 and 14)
1. Quickening
   Vernix caseosa
   Lanugo hair
   Foramen ovale
   Umbilical vein
   Ductus venosus

Parturition and Lactation (Objectives 15-21)
1. Parturition
   Labor
   Progesterone
   Estrogen
   Posterior pituitary gland
   Prostaglandins
   Positive feedback
   Dilation stage
   Expulsion stage
   Dilation stage
   Placental stage
   Cephalic (head first)
2. 1) Collapsed
   2) Umbilical cord
   3) Carbon dioxide
   4) pH (acidosis)
   5) Decreasing
   6) Medulla
3. Lactation
   Prolactin
   Oxytocin
   Estrogen
   Progesterone
   Colostrum
4. 1) Hypothalamus
   2) Prolactin releasing hormone
   3) Prolactin
   4) Milk production
   5) Ceases
   6) Days

Postnatal Development (Objective 22)
1. Neonatal
   Senescence
   Childhood
   Infancy
   Adolescence
   Infancy
   Childhood
   Senescence

☞ **Answers to Chapter Self-Quiz**

1. **Capacitation:** weakening of acrosomal membrane.
   Zygote: fertilized egg.
   Cleavage: mitotic cell divisions that occur after the zygote is formed
   Chorionic villi: projections of chorion that penetrate the endometrium.
   Senescence: period of old age; ends in death.
2. d
3. Ectoderm
   Mesoderm
   Endoderm
   Mesoderm
   Ectoderm
4. Chorion
   Allantois
   Yolk sac
   Amnion
   Yolk sac
5. b
6. c
7. c
8. b
9. d
10. c

☞ **Answers to Terminology Exercises**

Morphogenesis
Neonatology
Amniocentesis
Oxytocin
Morula

Before birth
False pregnancy
Difficult or painful labor
Process of giving birth
Old age

D (first pregnancy)
B (excessive vomiting during pregnancy)
A (cessation of milk secretion)
C (excessive quantity of amniotic fluid)
E (joining together of two cells)

☞ **Answers to Fun and Games**

ZYGOTE
CLEAVAGE
MORULA
TROPHOBLAST
AMNION
CHORION
PLACENTA

ORGANOGENESIS
LANUGO
VERNIX CASEOSA
PARTURITION
FORAMEN OVALE
OXYTOCIN
COLOSTRUM
NEONATE
ADOLESCENCE
SENESCENCE
TERATOGEN

☞ **Quiz/Test Questions**

Note: There are fifty multiple-choice questions for this chapter in the computerized test bank.

Name the following:

1. Region where fertilization generally occurs.
   Answer: uterine tube.

2. Single cell that is the result of fertilization.
   Answer: zygote.

3. Series of mitotic cell divisions that begin shortly after fertilization.
   Answer: cleavage.

4. Cells that enclose a blastocele.
   Answer: trophoblast.

5. Region of the blastocyst that develops into the embryo.
   Answer: inner cell mass.

6. Extraembryonic membrane that contributes to the formation of the placenta.
   Answer: chorion.

7. Period of prenatal development during which organogenesis occurs.
   Answer: embryonic period.

8. Period of prenatal development that lasts about 30 weeks.
   Answer: fetal period.

9. Vessels that carry fetal blood to the placenta for gaseous exchange.
   Answer: umbilical arteries.

10. Hair that covers the fetal body and helps keep the vernix caseosa in place.
    Answer: lanugo hair.

11. Longest stage of labor.
    Answer: dilation stage.

12. With prostaglandins, the hormone that stimulates uterine contractions.
    Answer: oxytocin.

13. Yellowish fluid that is secreted by the mammary glands prior to milk production.
    Answer: colostrum.

14. Postnatal period that begins at birth and lasts about 4 weeks.
Answer: neonatal period.

15. Period of old age that ends in death.
Answer: senescence.

True/False Questions:

1. For fertilization to take place, sexual intercourse must occur approximately between 1 day before and 3 days after ovulation.
Answer: False; intercourse should occur between 3 days before and 1 day after ovulation.

2. The second meiotic division in oogenesis is not completed until after a spermatozoan has penetrated the cell.
Answer: True.

3. The embryonic period lasts for about six months.
Answer: False; it lasts six weeks, from the end of the second week to the end of the eighth week after conception.

4. Trophoblast cells secrete human chorionic gonadotropin, which has an action similar to luteinizing hormone.
Answer: True.

5. The ectoderm, mesoderm, and endoderm develop during the first two weeks after fertilization.
Answer: True.

6. The amnion, chorion, and placenta form during the preimplantation period.
Answer: False; the extraembryonic membranes form during the embryonic period.

7. The expulsion stage of labor generally lasts for 3-4 hours.
Answer: False; it usually lasts less than one hour.

8. An umbilical vein carries oxygenated blood from the placenta to the fetus.
Answer: True.

9. Milk is not produced unless prolactin is stimulated by a baby's suckling.
Answer: True.

10. Temperature regulating mechanisms are well developed in the neonatal period.
Answer: False; they are not developed and the neonate is vulnerable to environmental temperatures.

☞ Answers to Final Forty

The "odd" item for each of the final forty is given below. The theme that links the four "even" items is indicated in parentheses.

1. Cranial cavity (ventral cavity)
2. Popliteal (ventral surface)
3. pH = 8 (acids)
4. Nitrogen (carbohydrates)
5. Nucleolus (organelles in cytoplasm)
6. Lysosomes (protein synthesis)
7. Smooth muscle (connective tissue)
8. Meninges (epithelial membranes)
9. Papillary (epidermal layers)
10. Sudoriferous (sebaceous glands)
11. Clavicle (axial skeleton)
12. Suture (freely movable joints)
13. Lactic acid (energy for contraction)
14. Gracilis (muscles of upper body)
15. Microglia (neurons)
16. Cranial nerves (central nervous system)
17. Pupil (refracting elements of eye)
18. Utricle (associated with hearing)
19. ADH (anterior pituitary gland)
20. Epinephrine (adrenal cortex)
21. $5,000/mm^3$ (RBCs)
22. Gamma globulin (blood clotting)
23. Mitral valve (associated with right atrium)
24. Ventricular filling (ventricular systole)
25. Femoral vein (oxygenated blood)
26. Renal artery (celiac artery and branches)
27. Right side of the face (thoracic duct)
28. Antiserum (active immunity)
29. Nasopharynx (lower respiratory tract)
30. Oxyhemoglobin (carbon dioxide transport)
31. Enterokinase (stomach secretions)
32. Lipase (act on carbohydrates)
33. Glycolysis (aerobic events)
34. Vitamin C (fat soluble vitamins)
35. Nephron loop (cortex of kidney)
36. Potassium ions (ECF)
37. Ureter (male reproductive tract)
38. Primary follicle (mature ovarian follicle)
39. Chorion (preembryonic period)
40. Fetal period (postnatal periods)

Chapter 1 Questions, *continued from page 5.*

11. Condition of a constant internal environment; the internal environment stays within normal ranges.
Answer: homeostasis.

12. Body cavity that contains the heart, stomach, liver, and urinary bladder.
Answer: ventral body cavity.

13. Abdominal region that is superior to the umbilical region.
Answer: epigastric.

14. Specific body cavity that contains the heart and lungs.
Answer: thoracic cavity.

15. Central abdominal region.
Answer: umbilical.

True/False Questions:

1. The cell is the simplest living unit of organization within the human body.
Answer: True.

2. The body system that transports oxygen to body cells is the respiratory system.
Answer: False; blood in the cardiovascular system transports oxygen.

3. The part of metabolism that synthesizes large molecules from smaller ones is called catabolism.
Answer: False; anabolism is the building-up process.

4. Heat and pressure are physical factors that are necessary for life.
Answer: True.

5. When your body needs water, you get thirsty, then you get a drink. This is an example of a positive feedback mechanism because you take a positive action by getting a drink.
Answer: False; it is negative feedback because taking a drink removes the stimulus.

6. In anatomic position your arms are at your sides and palms are facing forward.
Answer: True.

7. If you are standing on your head, your eyes are inferior to your mouth.
Answer: False; remember that directional terms refer to anatomic position.

8. The longitudinal plane that divides the body or an organ into anterior and posterior regions is the sagittal plane.
Answer: False; it is the frontal, or coronal, plane.

9. The most inferior abdominal region on the left side is the left lumbar region.

Answer: False; it is the left inguinal, or iliac, region.

10. The brachium and popliteal areas are part of the axial portion of the body.
Answer: False; they are part of the appendicular portion of the body.

Chapter 2 Questions, *continued from page 11.*

9. Type of reaction represented by this equation: $Cl_2 + 2NaBr \rightarrow 2NaCl + Br_2$.
Answer: single replacement or single displacement.

10. Substance that increases the rate of a chemical reaction.
Answer: catalyst.

11. Type of reaction in which more energy is stored in the products than in the reactants.
Answer: endergonic.

12. Substance that is dissolved in a solution.
Answer: solute.

13. Type of reaction in which an acid reacts with a base to produce a salt.
Answer: neutralization.

14. Type of organic compounds that contain nitrogen in addition to carbon, hydrogen, and oxygen.
Answer: proteins.

15. High energy compound that supplied energy in a form that is usable by body cells.
Answer: adenosine triphosphate (ATP).

True/False Questions:

1. The chemical symbol for calcium is C.
Answer: False; the symbol for calcium is Ca.

2. An element with 11 protons, 11 electrons, and 12 neutrons has a mass number = 34 amu.
Answer: False; mass number = protons + neutrons = 23.

3. An iron atom that loses three electrons becomes a positively charged ion.
Answer: True; when it loses three negative charges it becomes positively charged.

4. In the covalent molecule of water, the shared electrons are more closely associated with the oxygen than with the hydrogen. This makes water a polar molecule with the hydrogen end positively charged and the oxygen end negatively charged.
Answer: True.

5. If a base is added to a solution, the pH of the solution decreases.

Answer: False; a base increases the pH of a solution.

6. Crushing a substance into a fine powder tends to increase its reaction rate.
Answer: True.

7. A mixture of salt and water that is clear and does not settle is called a suspension.
Answer: False; it is a solution.

8. Fatty acids are the building blocks of proteins.
Answer: False; amino acids are the building blocks of proteins.

9. Glycogen and glucose are examples of carbohydrates.
Answer: True.

10. A sugar, a phosphate, and a nitrogenous base make up a nucleotide, which is the building block of a nucleic acid.
Answer: True.

Chapter 3 Questions, *continued from page 17.*

6. A poison that inhibits mitochondrial function will also inhibit or halt active transport processes in the cell.
Answer: True; mitochondria function in ATP production and without ATP active transport ceases.

7. A substance, such as glucose, that passes through the cell membrane by facilitated diffusion, requires a carrier molecule and energy from ATP.
Answer: False; facilitated diffusion requires a carrier, but does not require energy.

8. When a cell with an intracellular glucose concentration of 5 percent is placed in a 10 percent glucose solution, glucose molecules move from the 10 percent extracellular solution into the 5 percent intracellular solution by osmosis.
Answer: False; water moves out of the cell by osmosis.

9. DNA in the nucleus acts as a template for synthesizing mRNA.
Answer: True.

10. A cell that normally has 46 chromosomes undergoes mitosis and cytokinesis to produce two cells, each with 23 chromosomes.
Answer: False; after mitosis and cytokinesis, each of the two daughter cells will have 46 chromosomes.

Chapter 4 Questions, *continued from page 23.*

13. Structural unit of bone.
Answer: osteon of Haversian system.

14. Muscle tissue that is striated and involuntary.
Answer: cardiac muscle.

15. Intercellular material of blood.
Answer: plasma.

True/False Questions:

1. Actin and myosin are proteins found in nervous tissue.
Answer: False; actin and myosin are contractile proteins in muscle tissue.

2. Visceral muscle fibers are cylindrical and multinucleated.
Answer: False; skeletal muscle fibers are cylindrical and multinucleated.

3. Neurons are the conducting cells of the nervous system.
Answer: True.

4. Mucous membranes and serous membranes are epithelial membranes.
Answer: True.

5. The layer of a serous membrane that lines a body cavity is the visceral layer.
Answer: False; it is the parietal layer.

6. The innermost layer of the meninges is the dura mater.
Answer: False; it is the outer layer.

7. Epithelial tissues repair quickly because they have abundant blood vessels and reproduce readily.
Answer: False; epithelial tissues are avascular.

8. Microvilli are often found on columnar epithelium where absorption takes place.
Answer: True.

9. Another name for platelet is thrombocyte.
Answer: True.

10. The pleura, peritoneum, and pericardium are mucous membranes.
Answer: False; they are serous membranes.

Chapter 5 Questions, *continued from page 28.*

8. Type of holocrine gland that opens into hair follicles.
Answer: sebaceous gland.

9. Fold of stratum corneum that grows over the proximal portion of the nail body.
Answer: eponychium or cuticle.

10. Vitamin that is synthesized in the skin.
Answer: Vitamin D

11. Type of burn that causes reddened and blistered skin and is painful.
Answer: second degree burn

12. Two ways in which the skin helps to lower body temperature.
Answer: increasing activity of the sweat glands and dilation of the cutaneous blood vessels.

13. Epidermal layer next to the dermis.
Answer: stratum basale.

14. Integumentary glands that increase activity and decrease activity in old age.
Answer: sebaceous glands.

15. Portion of the hair that extends beyond the epidermis.
Answer: shaft.

True/False Questions:

1. Hair and nails are derived from the dermis of the skin.
Answer: False; they are derived from the epidermis.

2. The most numerous sweat glands are the apocrine sweat glands.
Answer: False; the most numerous are the merocrine glands that secrete perspiration.

3. Third degree burns are very painful.
Answer: False; they are not painful because the sensory receptors have been destroyed.

4. When Mr. Que received second degree burns to his entire right arm in a freak accident, he damaged 18 percent of his body surface area.
Answer: False; the arm constitutes 9 percent of the surface area.

5. Keratin is a protein in the skin that helps prevent water loss.
Answer: True.

6. The outermost layer of keratinized cells on the shaft of a hair is the cuticle.
Answer: True.

7. The dermis of the skin is highly vascular and contains glands.
Answer: True.

8. The integumentary system synthesizes melanin, keratin, and vitamin E.
Answer: False; it synthesizes vitamin D.

9. Older people tend to be more susceptible to temperature changes because they have less adipose in the subcutaneous layer.
Answer: True.

10. In most individuals, the epidermis is thicker than the dermis.
Answer: False; the dermis is thicker than the epidermis.

Chapter 6 Questions, *continued from page 36.*

5. A bone can increase in length as long as there is red marrow present.
Answer: False; the epiphyseal plate is the determining factor in bone growth.

6. There are 12 pair of true ribs.
Answer: False; there are 7 pair of true, or vertebrosternal, ribs.

7. Because of the ribs and vertebral column, there are more bones in the axial skeleton than in the appendicular skeleton.
Answer: False; axial skeleton has 80 bones and appendicular skeleton has 126 bones.

8. The fibula is a slender bone on the lateral side of the leg and is often fractured in skiing accidents.
Answer: True.

9. The joints between the vertebrae are classified as diarthrotic joints but their movement is limited by the muscles and ligaments along the vertebral column.
Answer: False; they are classified as amphiarthrotic, or slightly movable, joints.

10. The true pelvis is wider and more oval in females than it is in males.
Answer: True.

Chapter 7 Questions, *continued from page 43.*

5. Neurotransmitter at the neuromuscular junction.
Answer: acetylcholine.

6. A single neuron and all the muscle fibers it innervates.
Answer: motor unit.

7. Type of muscle contraction in which a muscle shortens and movement is produced.
Answer: isotonic.

8. Molecule within a muscle cell that replenishes the ATP used in contraction.
Answer: creatine phosphate.

9. Muscle that has an action that is opposite that of a prime move.
Answer: antagonist.

10. Type of movement at the elbow when you place your fingers on your shoulder.
    Answer: flexion.

11. Type of movement between leg and foot when you stand on tiptoes.
    Answer: plantar flexion.

12. Two muscles of mastication.
    Answer: masseter and temporalis.

13. Muscle that is the prime mover when you extend your forearm to receive a gift.
    Answer: triceps brachii.

14. Superficial chest muscle that adducts the brachium.
    Answer: pectoralis major.

15. Two muscles that form the calf of the leg.
    Answer: gastrocnemius and soleus.

True/False Questions:

1. Muscles are attached to bones by tendons.
   Answer: True.

2. The more stable end of a muscle is the insertion.
   Answer: False; the more movable end is the insertion.

3. Actin and myosin shorten, or contract, when a muscle contracts.
   Answer: False; the actin and myosin slide over each other to shorten the sarcomere.

4. The A band on a muscle fiber becomes shorter when the fiber contracts.
   Answer: False; the A band remains the same length, the I band shortens as the Z lines draw closer together.

5. A stimulus that exceeds threshold causes a greater than normal contraction of a muscle fiber.
   Answer: False; muscle fibers obey the all-or-none principle.

6. The long term energy source for endurance exercise, such as a marathon, comes from glucose and fatty acids.
   Answer: True.

7. Short periods of strenuous exercise that require anaerobic mechanisms to replenish ATP result in an accumulation of pyruvic acid molecules, which may lead to muscle soreness.
   Answer: False; it is lactic acid that accumulates.

8. The primary action at the knee when you kick a ball comes from the hamstring muscles.
   Answer: False; the primary action is extension which comes from the quadriceps femoris muscle group.

9. The sternocleidomastoid muscle allows you to flex your head and lower your chin to your chest.
   Answer: True.

10. Muscles that compress the abdominal wall include the external oblique and internal oblique.
    Answer: True.

Chapter 8 Questions, *continued from page 52.*

9. Lobe of the cerebrum that contains the primary visual area.
   Answer: occipital lobe.

10. Ventricle that is in the region of the diencephalon.
    Answer: third ventricle.

11. Region of the brainstem that contains the pneumotaxic and apneustic centers.
    Answer: pons.

12. Fluid channel between the third and fourth ventricles.
    Answer: cerebral aqueduct or aqueduct of Sylvius.

13. Cranial nerve that permits you to smile or frown.
    Answer: facial nerve (VII).

14. Tenth cranial nerve.
    Answer: vagus nerve.

15. Spinal nerve plexus that supplies innervation to the upper extremity.
    Answer: brachial plexus.

True/False Questions:

1. The neurilemma is a white fatty substance around the axons of some neurons.
   Answer: False; the white fatty substance is myelin.

2. The most abundant type of neurons is the bipolar neuron..
   Answer: False; the most abundant type is multipolar.

3. In inhibitory transmission at the synapse, the neurotransmitter hyperpolarizes the postsynaptic membrane and makes it more difficult to initiate an impulse.
   Answer: True.

4. Normally, cerebrospinal fluid is located in the subdural space.

Answer: False; CSF is normally in the subarachnoid space.

5. There is more white matter than gray matter in the cerebral cortex.
   Answer: False; the cerebral cortex is entirely gray matter.

6. Collectively, the commissural white fibers that connect the two cerebral hemispheres are called the corpus callosum.
   Answer: True.

7. The midbrain consists of the thalamus and hypothalamus.
   Answer: False; it consists of the cerebral peduncles and corpora quadrigemina.

8. In most people the spinal cord is at least 2 feet long.
   Answer: False; it is normally 17-18 inches long.

9. Somatic efferent pathways utilize two neurons, but the visceral efferent of autonomic pathway uses only one neuron.
   Answer: False; somatic efferent pathway uses one neuron and autonomic pathway uses two neurons.

10. The sympathetic division of the autonomic nervous system stimulates increases in heart rate, respiratory rate, and blood pressure.
    Answer: True.

Chapter 10 Questions, *continued from page 66.*

4. Mr. Que complains to his physician that he urinates frequently with a large volume of urine. His physician is likely to suspect a dysfunctional adenohypophysis resulting in hyposecretion.
   Answer: False; hyposecretion of the neurohypophysis results in large quantities of dilute urine.

5. Hormones from the parathyroid gland result in water retention and edema.
   Answer: False; parathyroid hormone elevates calcium levels in the blood.

6. Thyroxine levels are regulated by a negative feedback mechanism involving the hypothalamus and adenohypophysis.
   Answer: True.

7. A hormone from the thymus gland has an important role in the body's immune mechanism.
   Answer: True.

8. Calcitonin acts to increase blood calcium levels by absorbing calcium from the digestive tract.

Answer: False; calcitonin reduces blood calcium levels, parathyroid hormone increases the levels.

9. Aldosterone is an adrenal cortical hormone that is essential to life because it has a vital role in regulating the amount of sodium in the body.
   Answer: True.

10. Cells in the stomach and small intestine secrete hormones that regulate certain digestive activities.
    Answer: True.

Chapter 14 Questions, *continued from page 90.*

4. The thymus generally enlarges after puberty to provide additional protection for the adolescent.
   Answer: False; the thymus generally regresses after puberty.

5. A molecule, generally a protein, that triggers an immune response is an antigen.
   Answer: True.

6. B-cells are responsible for cell-mediated immunity.
   Answer: False; T-cells are responsible for cell-mediated immunity.

7. A secondary response relies on memory cells.
   Answer: True.

8. Immunoglobulins are antibodies.
   Answer: True.

9. Active immunity last longer than passive immunity.
   Answer: True.

10. Vaccines are used to provide passive artificial immunity.
    Answer: False; vaccines provide active artificial immunity and antisera provide passive artificial immunity.

Chapter 15 Questions, *continued from page 97.*

9. Inspiratory capacity is greater than vital capacity.
   Answer: False; vital capacity = TV + IRV + ERV and inspiratory capacity = TV + IRV.

10. Increased carbon dioxide concentration favors oxygen unloading.
    Answer: True.